MILITARY AND VETERAN ISSUES

PROFESSIONAL MILITARY EDUCATION

ANALYSIS AND RECOMMENDATIONS

MILITARY AND VETERAN ISSUES

Additional books in this series can be found on Nova's website under the Series tab.

Additional E-books in this series can be found on Nova's website under the E-book tab.

MILITARY AND VETERAN ISSUES

PROFESSIONAL MILITARY EDUCATION

ANALYSIS AND RECOMMENDATIONS

MICHELLE P. JACOBS
EDITOR

nova publishers
New York

For permission to use material from this book please contact us:
Telephone 631-231-7269; Fax 631-231-8175
Web Site: http://www.novapublishers.com

Library of Congress Cataloging-in-Publication Data

ISBN: 978-1-62948-814-1

Published by Nova Science Publishers, Inc. † New York

CONTENTS

PREFACE

The authors discuss professional military education two decades after the Goldwater-Nichols Act and the Skelton Panel and provide statements from important hearings on continued engagement.

Chapter 1 - This report examines officer in-residence professional military education (PME) as a critical investment in the most important element of our military – people. The primary purpose of PME is to develop military officers, throughout their careers, for the rigorous intellectual demands of complex contingencies and major conflicts. The United States cannot afford to be complacent when it comes to producing leaders capable of meeting significant challenges, whether at the tactical, operational, or strategic levels of warfare. Military officers must think critically, communicate well, conduct themselves with integrity, and lead others to perform strenuous tasks in difficult and often dangerous situations. As a matter of national security, the country's continuing investment in the PME system must be wisely made.

In supporting the military, the Congress is responsible for providing funds, setting associated policy, and providing oversight to ensure that all military and Department of Defense civilian personnel are properly prepared to perform their missions. The House Armed Services Committee has long supported the members of the armed forces by providing oversight, guidance, and resources with respect to PME. The most notable effort was the landmark review conducted by Chairman Ike Skelton's panel twenty years ago, which recommended comprehensive reform of the PME system.[1] That Panel's report stated: "Although many of its individual courses, programs, and faculties are excellent, the existing PME system must be improved to meet the needs of the modern profession at arms."

While this Subcommittee will not propose revolutionary changes as the Skelton Panel did, the current PME system should be improved to meet the country's needs of today and tomorrow. Twenty years ago, the U.S. military was educating officers to engage Cold War adversaries. Clearly, much about our military and our world has changed since then, and we know that much will continue to change as we look to the future. PME, therefore, must remain dynamic. It must respond to present needs and consistently anticipate those of the future. It must continuously evolve in order to imbue service members with the intellectual agility to assume expanded roles and to perform new missions in an ever dynamic and increasingly complicated security environment. Other requirements are enduring and must be preserved.

With respect to PME, Congress should regularly pose and assess these questions: How well is the nation educating its officers presently? And, what should be done to educate them more effectively in the future?

Chapter 2 - This is an edited, reformatted and augmented version of a Statement of Brigadier General Sean B. Macfarland, Deputy Commandant, Army Command and General Staff College, presented November 30, 2010 before the House Armed Services Committee, Oversight and Investigations Subcommittee.

Chapter 3 - This is an edited, reformatted and augmented version of a statement of Scott Lutterloh, Director, Total Force Training and Education Division, U.S. Navy, presented November 30, 2010 before the House Armed Services Committee, Oversight and Investigations Subcommittee.

Chapter 4 - This is an edited, reformatted and augmented version of a statement of Daniel R. Sitterly, Director of Force Development, U.S. Air Force, presented November 30, 2010 before the House Armed Services Committee, Oversight and Investigations Subcommittee.

Chapter 5 - This is an edited, reformatted and augmented version of a statement of Major General Robert Neller, President, Marine Corps University, presented November 30, 2010 before the House Armed Services Committee, Oversight and Investigations Subcommittee.

In: Professional Military Education
Editor: Michelle P. Jacobs

ISBN: 978-1-62948-814-1

Chapter 1

ANOTHER CROSSROADS? PROFESSIONAL MILITARY EDUCATION TWO DECADES AFTER THE GOLDWATER-NICHOLS ACT AND THE SKELTON PANEL*

House Armed Service Committee, Subcommittee on Oversight and Investigations

PREFACE

"The society that separates its scholars from its warriors will have its thinking done by cowards and its fighting done by fools."

Thucydides

This report examines officer in-residence professional military education (PME) as a critical investment in the most important element of our military – people. The primary purpose of PME is to develop military officers, throughout their careers, for the rigorous intellectual demands of complex contingencies and major conflicts. The United States cannot afford to be complacent when it comes to producing leaders capable of meeting significant challenges, whether at the tactical, operational, or strategic levels of warfare. Military officers must think critically, communicate well, conduct themselves with integrity, and lead others to perform strenuous tasks in difficult and often dangerous situations. As a matter of national security, the country's continuing investment in the PME system must be wisely made.

In supporting the military, the Congress is responsible for providing funds, setting associated policy, and providing oversight to ensure that all military and Department of Defense civilian personnel are properly prepared to perform their missions. The House Armed Services Committee has long supported the members of the armed forces by providing oversight, guidance, and resources with respect to PME. The most notable effort was the

* This is an edited, reformatted and augmented version of the House Armed Service Committee, Subcommittee on Oversight and Investigations, dated April 2010.

landmark review conducted by Chairman Ike Skelton's panel twenty years ago, which recommended comprehensive reform of the PME system.[1] That Panel's report stated: "Although many of its individual courses, programs, and faculties are excellent, the existing PME system must be improved to meet the needs of the modern profession at arms."

While this Subcommittee will not propose revolutionary changes as the Skelton Panel did, the current PME system should be improved to meet the country's needs of today and tomorrow. Twenty years ago, the U.S. military was educating officers to engage Cold War adversaries. Clearly, much about our military and our world has changed since then, and we know that much will continue to change as we look to the future. PME, therefore, must remain dynamic. It must respond to present needs and consistently anticipate those of the future. It must continuously evolve in order to imbue service members with the intellectual agility to assume expanded roles and to perform new missions in an ever dynamic and increasingly complicated security environment. Other requirements are enduring and must be preserved.

With respect to PME, Congress should regularly pose and assess these questions: How well is the nation educating its officers presently? And, what should be done to educate them more effectively in the future?

EXECUTIVE SUMMARY

The Subcommittee on Oversight and Investigations (the Subcommittee) reviewed the state of the officer in-residence professional military education (PME) system to determine what can, and should, be done to improve PME amid complex and evolving national security challenges. Military officers of every grade are expected to demonstrate intellectual agility, think critically, communicate well, conduct themselves with integrity, and lead others to perform strenuous tasks in difficult and often dangerous situations. The principal purpose of PME is to educate officers throughout their careers in preparation for this unique public trust. The Subcommittee endeavored to: evaluate PME's effectiveness relative to its purpose; assess whether it is sufficiently responsive to military needs; and appraise its component schools in their pursuits of well-resourced and qualitatively-rigorous programs. As a result, the Subcommittee identified specific areas for departmental action and further congressional oversight to promote continuing improvement of the system.

In 1987, the year following the passage of the Goldwater-Nichols Act[2] (Goldwater-Nichols), the House Armed Services Committee (the Committee) established a panel on PME led by Representative Ike Skelton (the Skelton Panel). The Skelton Panel undertook the last comprehensive congressional review of PME. The Skelton Panel assessed the PME system's ability to develop officers in both strategy and joint matters and the overall quality of PME, as well as the Department of Defense's (the Department's) plans to implement the joint PME requirements created by Goldwater-Nichols. The Panel published its findings and recommendations in a report, dated April 21, 1989 (the Skelton Report).

This Subcommittee did not attempt to reproduce either the scope or the depth of the Skelton Panel's historic review. Instead, the Subcommittee focused on those developments since the Skelton Panel's review that influence the mission effectiveness of the PME system and used the Skelton Report as a baseline. Since the Skelton Report, Congress has passed numerous pieces of legislation which affect the PME system both directly and indirectly. Congress has not comprehensively studied the accumulated effects on PME of 20 years of legislative changes. This report is only able to highlight a few areas that may deserve further examination. The Subcommittee finds that PME deserves more constant and frequent congressional oversight.

The Subcommittee's examination of PME was conducted mindful of PME's contributions to the wider purposes of officer development. PME is a component of the developmental process that includes training, experience, and self-development. The Subcommittee considers PME to be critical because it empowers individual improvement through thinking and learning, which collectively ensures institutional growth.

PME encompasses a diversity of subject matter. Each service is responsible for educating officers in their core competencies according to service needs. Air Force schools, for example, primarily teach air and space warfare. Similarly, Army, Navy, and Marine Corps schools focus on land, maritime, and expeditionary warfare, respectively. The Department depends on the services' PME to develop officers with these service-specific proficiencies.

Goldwater-Nichols recognized that, in addition to service-competent officers, the Armed Forces need high-quality officers competent in joint matters. In drafting the Act, Congress envisioned the development of this latter group of officers through a combination of joint professional military education (JPME) and joint duty assignments. At first, the distinction between, and the delivery of, PME and JPME were more clearly defined. Originally, an officer needed to complete JPME to become a joint specialist. That JPME was only available at three joint schools and only a small group of specialists was believed to be needed. Now, JPME is taught outside the original three joint schools and is included within the curricula of the services' war colleges. Where it had been more distinct from service-specific PME in the past, JPME now permeates the services' traditional PME. PME and JPME, together, prepare officers in successive stages throughout their careers to engage intellectual challenges appropriate to increases in their ranks and responsibilities. However, as an officer advances, he or she must elevate his or her service-specific proficiencies, while accumulating increasing amounts of expertise in joint matters. As a result, the proportion of JPME in an officer's professional education increases with seniority. In the past, JPME seemed to be more of a bridge from service competency to joint specialty for a select cadre. Today, traditional PME curricula would be considered incomplete without the inclusion of joint curricula. Consequently, this report uses the term PME to include JPME, unless otherwise specifically noted.

Like the Skelton Panel, the Subcommittee agrees that the Department's PME system is still basically sound. However, there are areas of PME that need improving. As a means for facilitating improvement, the Subcommittee offers two sets of findings and recommendations: those that concentrate on systemic issues; and those that are specifically intended for individual schools, and their leaders, faculty members, and students. The latter are termed "institutional issues" for the purposes of this report.

The first set of findings and recommendations, *system issues*, are described in this summary in the first four paragraphs below, while those in the second set, *institutional issues*, are described in the final three. In the list of major findings, the first eight are covered in the systems section and the last three address institutional issues.

First, the Subcommittee found that the PME system has been significantly modified over the past 20 years both in practice and in legislation, the results of which raise issues with the timing, purpose, and effectiveness of not only PME but also JPME. For example, two recent studies have indicated that officers are serving in joint and service staff assignments without adequate educational preparation. These studies have pointed to specific deficiencies in areas such as critical thinking that can and should be addressed throughout an officer's professional military education.

With regard to JPME specifically, the Department's implementation of recent legislative changes has weakened the connection between JPME and joint duty assignments. Yet, JPME completion is needed to be eligible for appointment to general or flag officer. The combination of these changes suggests that JPME completion may be more relevant to ensuring an officer's competitiveness for selection to flag rank than it is to enhancing job performance in the joint arena. The Subcommittee questions this approach and believes it warrants further examination. There is a tension between the officer's assignments necessary for career development, the needs of the joint force, and professional military education, whether it is at a military institution or a civilian institution.

Second, the Subcommittee found mixed results with respect to one of the most important areas that the Skelton Panel addressed – the cultivating of military strategists. Joint and service efforts are relatively disassociated from one another. Although PME is a factor in these efforts, it is not the primary means for shaping strategists. The Subcommittee recommends that the Joint Staff and each of the services carefully review and coordinate their PME efforts with the goal of educating qualified strategic decision-makers (in addition to strategic analysts and advisors) for service in positions of senior command authority. As part of that review, the Subcommittee recommends that the Joint Staff and services consider, in addition to PME, sponsoring additional junior officers for civilian masters' and doctoral degrees in strategy-related disciplines (e.g., history, political science, economics, international relations) at top-tier civilian universities. These officers should be provided command and staff assignments as well as positions on PME faculties, as appropriate.

Third, regarding the qualitative content and delivery of PME, the Subcommittee found that joint and service-specific PME curricula have evolved and rigor has improved since the Skelton Panel made its recommendations. It is a constant challenge for the Joint Staff and the services to balance enduring professional educational requirements with emerging operational needs. This challenge highlights the system's ability to adapt. PME curricula have adapted at differing, but generally appropriate, levels to new demands for instruction in language and culture; irregular warfare; and joint, interagency, intergovernmental, and multinational operations. Again, in differing but adequate measures the PME system retained suitable emphases on the enduring subjects of history and strategy. While recognizing that the services have organizations dedicated to looking ahead to doctrine that may be needed in the future, the Subcommittee found that many curricular developments were appropriately responsive to changing demands, but few, if any, of these developments were effected in anticipation of emerging opportunities and challenges. The Subcommittee believes that a more balanced approach to curriculum development throughout the PME system may succeed in fostering the sort of forward thinking associated with the Naval War College in the years preceding World War II, when the staff and students at the Naval War College were renowned for their wargaming. They successfully planned for the major Pacific War battles except Pearl Harbor and use of *kamikaze* pilots.[3]

Fourth, the Subcommittee surveyed many of the organizational, human, and material factors that contribute to the effective practice of PME. While the separate schools have improved considerably due to the Skelton Panel's recommendations, there are still some areas where greater improvement can be made. At the top level, organization matters. The Subcommittee found the following: (1) on both the joint and service levels, and even at the various schools, organizational structure should be examined for possible improvements that could enhance effectiveness; and (2) senior leaders should remain aware of the need to maintain "ownership" of professional military education and to support those involved in the PME enterprise.

As with any educational enterprise, there are also a number of human resource issues requiring attention. The Subcommittee found that: (1) more defined criteria for selection and longer tour lengths should be considered for the senior leaders of the schools to optimize their ability to contribute to the PME mission; (2) faculty can be better supported in a number of ways; (3) the composition of PME institution student bodies (military, interagency, international, and private sector) should be constantly reviewed to ensure that their characteristics support the schools' educational missions. The services and Department must improve the selection process by more carefully correlating prospective students' experience, qualifications, and likely subsequent assignments with selection to specific schools. Similar planning must be exercised in choosing the appropriate follow-on assignments for graduates. Finally, adequate material support is also necessary for success. While PME is not poorly resourced, some material challenges are apparent. These challenges include needs for reliable funding streams for the joint schools, renewed infrastructure for the older schools, and appropriate information and educational technology for many schools.

MAJOR FINDINGS

- Today's PME system is basically sound; there are areas, however, that need improvement. The system operates within a dynamic national security environment. Consequently, it must be more prepared to anticipate and adapt to current and future challenges.
- PME's overall conceptual design has changed significantly since 1989, particularly with respect to the educational content needed and offered to the most junior officers and at the war colleges. There is an increasing need for additional joint and service-specific subject matter to be taught earlier in officers' careers. PME has also broadened with respect to the teaching of strategy. Law and policy now require that national security strategy be taught at each of the senior PME schools.
- Competing demands make it difficult to accommodate the need for the requisite PME, training, and experience. Officers are finding it increasingly challenging to complete their required PME, which is only compounded by current operational requirements. This strain has contributed to the services seeking flexibility in managing PME and assignments.
- With limited exceptions, nothing in law now precludes officers from being assigned to joint billets without having received JPME credit. As a result, many officers are assigned to joint billets without having completed appropriate joint education. This disconnect between JPME and joint duty assignments has become a common

practice, disregarding a fundamental purpose of JPME, which by law and policy, is preparation for those assignments.

- Some operational commanders, including the Combatant Commanders, reportedly consider their staff officers lacking in certain critical abilities necessary to perform their jobs effectively. Significant numbers of officers are serving in staff positions without having appropriate levels of PME prior to assignment. Furthermore, many officers reportedly consider the PME they receive to be inadequate preparation for these assignments.
- Joint and service efforts to identify and cultivate strategists are disassociated from one another. Although officer in-residence PME is a factor in these efforts, it is not the primary means for developing future strategic decision-makers. All of the services should cultivate strategists to assume positions of senior command authority
- The balancing of enduring PME requirements with emerging needs presents continual challenges. The schools' curricula have adapted at differing, but generally appropriate, paces. However, these developments appear to have been in reaction to changing demands rather than in anticipation of them.
- PME institutions have generally implemented the Skelton Panel recommendations on improving teaching practices and have adopted more demanding standards. Student-centered seminar discussion groups are the core means of instruction at the in-residence schools. Although PME institutions have adopted a variety of practices with regard to grading, these practices do not necessarily detract from the rigor of the academic programs.
- Coordinated direction of PME is important: (1) leaders at every level (e.g. the Office of the Secretary of Defense, the Joint Chiefs of Staff, service, and school) must build and maintain a comprehensive awareness of PME matters and facilitate decision-making processes to promote PME mission effectiveness; and (2) senior leaders must continually reaffirm their support for PME and those involved in the PME enterprise.
- The Skelton Report focused attention on PME leaders, faculty, and students. People remain the most important element of PME. PME leaders, faculty, and students must be carefully selected for their responsibilities. The Subcommittee's findings include: (1) senior leaders' tour lengths do not necessarily recognize the complexity of PME institutions and allow for stability in the management of those institutions; (2) PME faculties could be better supported in a number of ways; and (3) the services use differing processes for selecting their own students and varying approaches for attracting international, interagency, other service, and industry students, producing disparate results with respect to the composition and the quality of PME student bodies.
- Adequate material support is also necessary for PME success. While PME is not poorly resourced, some material challenges are apparent. These challenges include needs for reliable funding for the joint schools, renovation for schools with older infrastructure, and appropriate information and educational technology for many schools.

INTRODUCTION

"Professional attainment, based upon prolonged study, and collective study at colleges, rank by rank, age by age— those are the title reeds of the commanders of future armies, and the secret of future victories."[4]

Winston Churchill, 1946

It is Congress' responsibility, through its authorization of funds and statements of policy, along with the leaders of the Department of Defense (the Department), to ensure that military personnel who are asked to support the national security of the United States are properly prepared and equipped for their missions.

A program for development leading to commissioning and continuing through the length of a career supports the preparation of military officers who lead the armed forces. For the most senior, those with the most responsibility, careers stretch from 20 to 40 years. Officer development programs include providing the right officers the right training, experience, and education at the right time. The principal purpose of professional military education (PME) is to educate officers throughout their careers in preparation for this unique public trust.

The U.S. Armed Forces generally recognize the value of education separate from training, and they place special emphasis on the importance of in-residence officer education. PME contributes to an officer being able to take on responsibilities and challenges commensurate with increases in rank. The services seek to instill competence in core service functions and specific weapon systems in their officer candidates and junior officers. This knowledge is to be broadened to the operational level (combined arms and joint campaigns) for majors and lieutenant colonels (Navy lieutenant commanders and commanders, O-4s and O-5s). Finally, the military requires policy and strategic-level thinking from its colonels and flag officers (O-6 through O-10). Generally, training programs are highly utilitarian while the education system, particularly at the senior level, is intended to develop habits of mind and modes of analysis. As many military leaders have said, "we train for certainty and we educate for uncertainty." Still, all of the PME courses have elements of both training and education. By and large, the more junior the officers, the heavier the component of training in the courses they take. The more senior the officers, the heavier the education component in their courses.

The Subcommittee on Oversight and Investigations (the Subcommittee) examination of PME was conducted mindful of PME's contributions to the wider purposes of officer development. PME is a component of that developmental process. The Subcommittee considers PME to be critical because it empowers individual improvement through thinking and learning, which collectively ensures and increases institutional effectiveness.

PME encompasses a diversity of subject matter. Each service is responsible for educating officers in their core competencies according to service needs. Air Force schools, for example, primarily teach air and space warfare. Similarly, Army, Navy, and Marine Corps schools focus on land, maritime, and expeditionary warfare, respectively. The Department depends on the services' PME to develop officers with these service-specific proficiencies.

As this report will describe, service-specific PME existed before joint education. And joint education existed before formal joint PME (JPME), which was established as a result of congressional action in the late 1980s. The Subcommittee acknowledges one assumption at the outset: Each of the components should maintain service PME institutions that are the centers of excellence in their respective warfare domains. In the National Defense Authorization Act for Fiscal Year 1994 Congress specified that the primary mission of service PME schools is to provide expertise in particular warfare specialties, while the primary mission of JPME is to provide expertise in the integrated employment of land, sea, and air forces and that both PME and JPME programs are necessary in the education of military officers.[5]

The Chairman of the Joint Chiefs of Staff's (CJCS) policy guidance for officer PME acknowledges that officer professional development is a service responsibility. The services are responsible for their academies and their primary courses (captains and Navy lieutenants, O-3s), intermediate-level education, and senior-level education, as well as component flag officer courses. The CJCS guidance describes JPME as "embedded" within the PME system and emphasizes that the Joint Staff is tasked with overseeing the JPME program. Consequently, the service PME schools at the intermediate-level award JPME I and are accredited for the JPME component of their curricula through a process known as the Program for the Assessment of Joint Education (PAJE). The service schools at the senior-level are subjected to accreditation through PAJE for delivery of JPME II.

Over the years, service and joint PME have become intertwined to some measure, which should become evident in the system described in the pages that follow. This is in part due to the services embracing joint operations to the point where multi-service cooperation has become the norm. Joint doctrine in many cases also serves as service doctrine. This assimilation has even extended down to the level of joint tactics, techniques, and procedures. This overlap also gives rise to confusion in discussing, and sometimes equating, PME and JPME.

This report, in its treatment of current issues, necessarily gives significant exposure to those issues related to JPME. In no small measure, this is a consequence of adjustments to the JPME system being the focus of legislation over the preceding two decades, while PME has seen less change. Additionally, it also reflects the growth of content of joint curricula which now encompasses subject matter on topics such as cyberwarfare, information operations, and strategic communications. The content of this report reflects both of these trends.

The CJCS is responsible for joint education at the National Defense University including JPME I and II ("single-phase") at National War College and the Industrial College of the Armed Forces; JPME II at the two schools under the Joint Forces Staff College – the Joint and Combined Warfighting School and the Joint Advanced Warfighting School; and, the Capstone course for flag officers (now JPME III).[6] Although the Capstone course and service pre-commissioning and primary PME courses are charged by the CJCS with delivering joint education, those programs currently do not receive accreditation through the PAJE process.

Purpose

Professional military education is the backbone in the development of the nation's armed forces, and the quality of that military education distinguishes U.S. forces around the world. The House Armed Services Committee (the Committee) remains committed to ensuring that the quality and availability of PME programs remain a priority for the services and the Department, even during times of high operational tempo when they may be tempted to shortchange investment in educational opportunities to provide manpower and resources to other efforts.

As part of its oversight responsibilities, the Subcommittee engaged in a review of the rigor and relevance of the curricula being offered at all levels of in-residence officer PME. We also sought to comprehend how well the PME system and institutions have adjusted to realities associated with 21st century geopolitical and technological change in four key areas of the curricula: joint and interagency integration; language skills and cultural awareness; irregular warfare and stability operations; and history. At the same time, the Subcommittee examined the context for the PME system. In other words, this report describes what PME contributes to the larger system of officer development and how the personnel system intersects with the PME system. Finally, we reviewed the organization of the system and institutions as well as the human and material resources dedicated to them.

The findings and recommendations of the 1988 House Armed Services Committee Panel on Professional Military Education (the Skelton Panel, named for Representative Ike Skelton) were this project's starting point.[7] The Skelton Panel made specific recommendations on how the Goldwater- Nichols Department of Defense Reorganization Act of 1986 (Goldwater-Nichols) "joint" reforms should be institutionalized among the services through officer in-residence education. This Subcommittee did not attempt to reproduce either the scope or the depth of the Skelton Panel's historic review. Instead, the Subcommittee focused on those developments since the Skelton Panel's review that influence the mission effectiveness of the PME system. Since the Skelton Report, Congress has passed numerous pieces of legislation that affect the PME system both directly and indirectly. The Subcommittee wanted to understand the extent to which the recommendations and legislative provisions of past Congresses have been implemented.

Despite those numerous pieces of legislation, Congress has not comprehensively studied the accumulated effects on PME of 20 years of legislative changes, and this report is only able to highlight a limited number of areas that may deserve further examination. This Subcommittee has identified specific areas for departmental action and further congressional oversight to promote continuing improvement of the system. This report discusses a number of issues in significant detail as part of Congress' oversight responsibility. However, the Subcommittee does not think it appropriate to legislate on most of these matters. Readers will find few legislative proposals among our recommendations. The Subcommittee recommends, at the very least, that PME is important enough to warrant more constant and frequent congressional oversight.

Scope

Although this study may spark interest in a top-to-bottom review of military training and education, it was beyond the Subcommittee's intent to address the ways in which military education institutions and training venues teach all of the competencies officers need to do their jobs effectively. For instance, this report does not delve deeply into all of the aspects of the service academies or occupational specialty schools. The extensive training systems of the services, combatant commands, and the Department are also beyond the scope of this project. In addition, military education in and for the Coast Guard and Merchant Marine, Reserve Officer Training Corps programs, Officer Candidate or Training Schools, the regional centers such as the Marshall Center, non-resident programs, enlisted PME, and comparisons to foreign PME systems are beyond the scope of this study. There are also many ways in which military members (officers and enlisted) and civilians seek professionalization from technical and operational training (individual and unit) and education that this study does not address. Also beyond the scope of this study are the Department's very important civilian professional education and development programs and programs for the military's reserve component, except as they bear on this report's larger focus on officer in-residence PME. Finally, this report touches on interagency students and faculty participation in and contributions to PME. However, the report does not discuss the larger issue of interagency professional education such as proposals for a national security university or the National Security Professional Development program. Although important, these are separate and distinct from PME.

While just addressing the schools that constitute the bulk of officer in-residence PME would be challenging enough, the Subcommittee realized that one could not look at the schools in a vacuum. The study would also have to review the Department, Joint Staff, and service systems for the administration and direction of both PME and JPME. Beyond this context, the Subcommittee sought to explore the larger purpose the PME system serves. In other words, the Subcommittee examined the legal and policy contexts of officer development within which the PME system exists. The Subcommittee had to look to some degree at where and how officers come to PME and to which assignments they go after they complete each program. Do the joint and service officer management and personnel systems capitalize on the investment the nation makes in these officers' education? The one certainty in this construct, as the Subcommittee discovered and this report explains, is the tension between service and joint education and assignment requirements.

As much as the Subcommittee would like to have addressed each of the myriad complexities surrounding the PME and the joint and service officer management and personnel systems, many questions remain. In some cases, people interviewed and those who testified found remarkable agreement on how to resolve challenges. In other cases, there was no consensus. The Subcommittee introduces some of them without analysis or judgment in a separate section at the end of this report under "Issues for Further Study."[8]

Approach

This report is divided into five sections of varying length. The first gives the background or context within which the PME institutions operate, including a brief history, estimates of the current and future security environment, and the Department's, CJCS', and the services' PME policies. The next two are the longest sections, which discuss the practical issues the Subcommittee focused on as well as observations, findings, and recommendations. Like the Skelton Panel, the Subcommittee agrees that the Department's PME system is still basically sound. However, there are areas that need improvement. As a means for facilitating improvement, the Subcommittee offers two sets of findings and recommendations: first, those that concentrate on systemic issues and, second, institutional issues or those related to organization and those specifically intended for individual schools, and their leaders, faculty, staff, and students. In this section, leadership and faculty are dealt with at some length. Like the Skelton Panel, this Subcommittee finds that leaders and instructors are the bedrock of the PME system. The fourth section briefly identifies a number of challenging areas of study that remain as well as some individuals whose broader proposals arose in the course of testimony, interviews, current debates, and recent writings. These bear further and more in-depth consideration than could be provided at this time. Finally, a short conclusion provides a nascent vision of the essential attributes of future officers. Professional military education must contribute to developing those attributes.

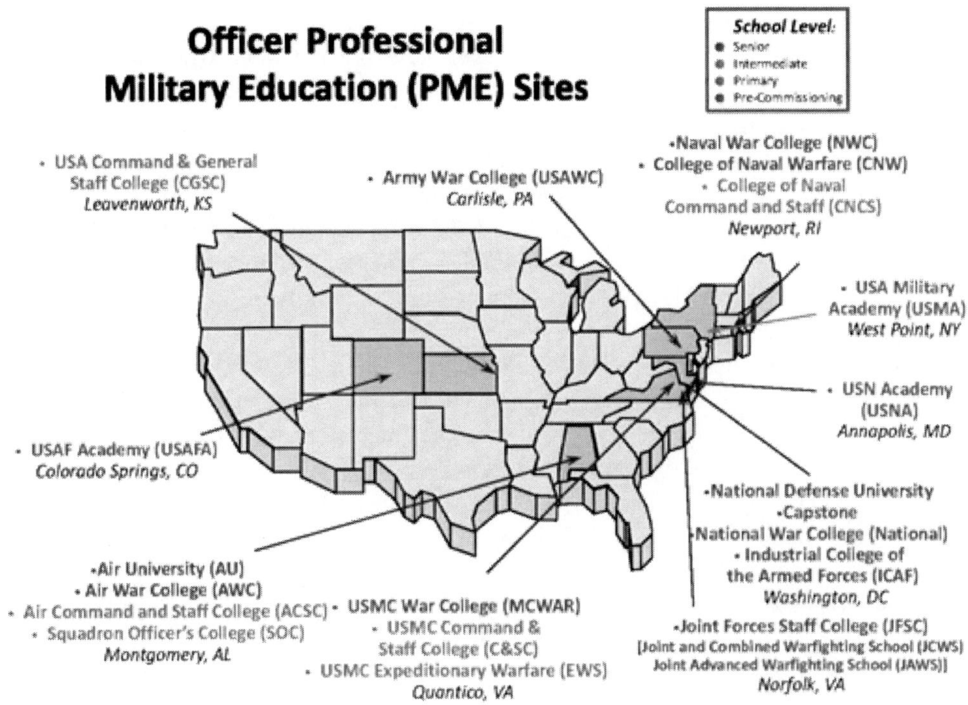

BACKGROUND

> *"...it falls to Congress to ensure our military strength is adequate to defend the nation and national interests. Indeed, there is no more important duty for Congress than to provide for the common defense."*[9]
>
> Representative Ike Skelton
> *Whispers of Warriors*, 2004

This section of the report starts with a brief history of professional military education (PME) in the United States before the Goldwater-Nichols Department of Defense Reorganization Act of 1986 (Goldwater-Nichols) and the immediate effects of that law and the resulting House Armed Services Committee Panel on Professional Military Education (the Skelton Panel). It then outlines the current and projected security environment, as well as this study's context – joint and service PME and personnel policy.

PME before the Goldwater-Nichols Act[10]

The United States, like most other nations, has reformed its professional military schools after conflicts inevitably revealed shortcomings in the performance of its armed forces. Lessons learned in wars generally have been preserved, refined, and inculcated throughout the services in an academic environment. The modern PME system had its beginnings in the early 19th century when major European states realized that they needed better educated militaries for large-scale, ideologically-motivated, industrialized wars. Officers had to study the essential dynamics of the art and science of war more formally. European pre-commissioning schools came first, and the United States joined this movement by establishing the United States Military Academy in 1802 and the United States Naval Academy in 1845. Although there was an almost universal belief that war is an intensely human endeavor and is bound up in social, cultural, economic, and political interactions, these academies were focused heavily on technical and engineering courses.[11] Post-graduate officers' courses started to be developed, modeled on the Prussian example, in this country after the Civil War.

By the beginning of the 20th century the basic PME framework had been established. The intermediate study of the art of war was conducted at the Army's Command and General Staff College (CGSC), evolving after 1881 from the Infantry and Cavalry School, and after 1884 at the Naval War College. Senior sea-service officers also studied at the Naval War College, while the Army established its war college in 1901 after the Spanish-American War.

The most extensive changes to PME in the United States occurred in and following World War II. Serious consideration was given to including more joint education for officers and synchronizing it with service PME. In 1943 the Army-Navy Staff College (ANSCOL) was established to provide a four-month course for select officers assigned to unified command and staff duties. In January 1946, the War Department commissioned a major study of officer education under the direction of Lieutenant General Leonard T. Gerow, Commandant of the CGSC. The February 1946 report of the "Gerow Board" recommended five joint colleges that would collectively form the "National Security University" located in Washington under the Joint Chiefs of Staff. The Industrial College of the Armed Forces (ICAF) already existed (established in 1924) to educate officers from all services in mobilization, supply, and industrial support. The Board proposed adding a national war college; a joint administrative (personnel and manpower) college; a joint intelligence college (today, the National Defense Intelligence College); and a Department of State college (today's Foreign Service Institute conducts short training courses but does not grant degrees). Some of the Gerow proposals were rejected because of resource limits. However, ICAF remained open and the Army War College which had suspended operations during World War II, reopened and moved to Carlisle Barracks, Pennsylvania. National War College (National) was established and took up residence in the former Army War College facilities in Roosevelt Hall on Fort Leslie J. McNair in southwest Washington, D.C. The Armed Forces Staff College (AFSC), which evolved from ANSCOL, moved to Norfolk, Virginia, in 1946 to provide joint operational instruction to mid-grade officers. The Navy retained its highly-regarded college in Newport, Rhode Island,[12] and the newly-established Air Force (1947) grew its war college from the former Army Air Corps Tactical School in Montgomery, Alabama. After the Vietnam War, the National Defense University (NDU) was established to consolidate management of the three joint schools (ICAF, National, and AFSC).

Assistant Secretary of the Navy Roosevelt (front center) at the Naval War College, c. 1897.

The Goldwater-Nichols Act

> *"Unity of Command and effort has been a cardinal principle of successful military organizations throughout history. Coaxing that coordinated effort out of the separate armed services with different cultures and command structures always has been a challenge for U.S. [government and] military leaders."*[13]
>
> Richard Cheney and Bill Taylor
>
> *Professional Military Education: An Asset for Peace and Progress*

Attempts to coordinate service efforts are not new. In recent times, the most significant of these has been the Goldwater-Nichols Act.[14] At that time, the Armed Services Committee (the Committee) broke the standard pattern of major military reforms being initiated by the White House. The Committee held hearings on flawed operations in the early 1980s including the attempt to rescue Americans held hostage in Iran and the invasion of Grenada. Goldwater-Nichols established clearer lines of command and control and improved the ability of the services to work with each other in truly joint, rather than simply multi-service, operations. It reduced the influence of the Service Chiefs operationally, even as they retained the responsibility to organize, train, and equip their forces. At the same time, Goldwater-Nichols increased the power of the Chairman of the Joint Chiefs of Staff (CJCS) and the Combatant Commanders (COCOMs). In order to strengthen interoperability and a commitment to joint operations, Congress mandated that positions on joint staffs and in joint commands would generally be filled by qualified joint specialty officers (JSOs) or those officers who were on track to become JSOs. JSOs had to complete a two-phased Program for Joint Education consisting of joint PME I (JPME I) and JPME II. To improve the quality of officers assigned to joint duty, once considered a career dead end, Congress also made joint duty a prerequisite for advancement to flag (general or admiral) rank.

Goldwater-Nichols recognized that, in addition to service-competent officers, the Armed Forces need high-quality officers competent in joint matters. In drafting the Act, Congress envisioned the development of this latter group of officers through a combination of JPME and joint assignments. At first, the distinction between, and the delivery of, PME and JPME were more clearly defined. Originally, an officer needed to complete JPME to become a joint specialist. JPME was available only at three joint schools, and only a small group of specialists was believed to be needed. Now, JPME is taught outside the original three joint schools and is included within the curricula of the services' command and staff and war colleges. Where it had been more distinct from service PME in the past, JPME is now integrated into the services' PME. PME and JPME, together, prepare officers in successive stages, throughout their careers, to engage intellectual challenges appropriate to increases in their ranks and responsibilities. However, as an officer advances, he or she must elevate his or her service-specific proficiencies, while accumulating increasing amounts of expertise in joint matters. As a result, the proportion of JPME in an officer's education increases with seniority. In the past, JPME seemed to be more of a bridge from service competency to a joint specialty in a select cadre. Today, service PME curricula would be considered incomplete without the inclusion of joint curricula. Consequently, this report uses the term PME to include JPME, unless otherwise noted.

THE SKELTON PANEL – RECOMMENDATIONS AND STATUS

The Skelton Panel was formed in the wake of Goldwater-Nichols, and it undertook a comprehensive review of PME. Its charter, signed by then Committee Chairman Les Aspin, called on the Panel to review joint education requirements under the Goldwater-Nichols legislation and assess the military's ability to develop military strategists, joint warfighters, and tacticians.[15] The Skelton Panel conducted an exhaustive review of the joint and service PME schools as well as several foreign military education institutions. The Panel's primary purpose was to review DOD plans for implementing Goldwater-Nichols JPME requirements, because a change was required in the service-centric mindset of military officers, and the Committee understood that a key way to change attitudes was through education.

The Panel's findings appeared in the 1989 *Report of the Panel on Military Education of the 100th Congress of the Committee on Armed Services* (the Skelton Report) and fell into two broad categories: the first established a conceptual model in which each level of education built on previous levels and each college or institution had a clear, fundamental teaching focus. The other urged restoring two joint colleges – National at the senior level and the Armed Forces Staff College, now the Joint Forces Staff College (JFSC), at the intermediate level – as centers of excellence and to the prominence they enjoyed in the early post-World War II period.[16] While recognizing that the successful officer first had to be an expert on his or her service's capabilities, the Panel envisioned the introduction of new joint concepts at the intermediate level (staff college) and the expansion on these concepts at the senior level (war college).[17]

The Skelton Panel recommended a two-phased joint education process that would be a subset of existing service PME. The intermediate service schools still had the primary function of educating officers in their respective warfare specialties, but they were also given a role in promoting joint education. The Skelton Panel saw a joint officer as having, "a thorough knowledge of his or her own service, some knowledge of the other services, experience operating with the other services, and the perspective to see the 'joint picture.'"[18] The Skelton Panel suggested that this was best accomplished at long, in-residence schools.[19] JPME I required familiarity with each service's doctrine, organizational concepts, and command and control.[20] In addition, students would be introduced to joint planning processes, joint systems, and the role played by service component commands in the unified command structure. For JPME I, the Skelton Panel recommended that the mix of students should be two officers per each seminar from each of the other two services. For faculty, the mix was 70 percent host and 15 percent each from the other services (instead CJCS policy established the mix for students as one non-host officer per service per seminar and for faculty as 90:5:5).[21]

The second phase of joint PME (JPME II) was to be delivered at the AFSC and would be an in-depth course of study in the integrated deployment and employment of multi-service forces. This course was to build on JPME I, be delivered when an officer was en route to a joint assignment, and classes would be only in-residence, multi-service, and on neutral ground in order to achieve joint acculturation or "socialization."[22] It would remain a 12-week course (more recently it was reduced to 10 weeks to increase "throughput"). CJCS policy empowered the two joint senior-level schools (National and ICAF) to deliver both the first and second phases of joint education in one 10-month period, along with their educational mission on operational art and strategy (and ICAF's mission of education on mobilization and resources).[23]

The services each retained their own senior school. Starting in 1989, these schools also awarded JPME I. They focused primarily on elements of joint warfare including component capabilities, operational art, and national military strategy. In response to service and DOD requests to increase throughput, Congress gave the Department the flexibility to award JPME I via intermediate-level distance education and to deliver JPME II at the service senior-level schools.[24] The latter are required to maintain at least a minimal mix of other service students and faculty (originally recommended by the Skelton Panel to be 50:25:25, but established by CJCS policy as no more than 60% host service).[25] In addition, the principle of "neutral ground" was abandoned, and the legal and policy mechanisms for assigning officers to JPME and joint assignments came to be perceived more as a path to promotion than as a requirement for conducting effective operations as a joint force.

Although PME is intended to serve a number of purposes, the study of strategy is critical to any discussion of officer education. The Skelton Panel was focused on developing strategists. Representative Ike Skelton has written that strategists are developed over a career and a lifetime, rather than only at senior-level PME. He cited General John R. Galvin as having written in 1989, "We need senior generals and admirals who can provide solid military advice to our political leadership…and we need young officers who can provide solid military advice, options, details, the results of analysis to the generals and admirals." Representative Skelton has also long believed that the study of history, particularly military history, is the key to developing leaders and strategists. He echoed others in writing, "It is a process of education, study, reading, and thinking that should continue throughout an entire military career. Yes, tactical proficiency is very important, but so too is strategic vision. That can only come after years of careful reading, study, reflection, and experience."[26]

Overall, the Skelton Panel made recommendations to the Department in nine areas:

- Establish a framework that specifies primary educational objectives at each level of PME – tactical level for pre-commissioning and primary (grades O-1–O-3), operational level for intermediate (O-4), and strategic level for senior (O-5–O-6) and flag officer levels (O-7–O-9).
- Establish a two-phase JPME program with JPME I at service colleges (intermediate and senior) and Phase II at AFSC.
- Focus senior service colleges on national military strategy and increase the other-service (non-host) faculty and student percentages at these schools (to 50:25:25). Focus National on national security strategy.
- Require intermediate and senior colleges to employ frequent graded essay exams and student reports.
- Determine whether the Navy should create more distinct curricula for its intermediate and senior schools.
- Convert National to a National Center for Strategic Studies and elevate the College to a level above the service colleges and ICAF (this was not implemented, but portions of the plan have become components of the current NDU).
- Require the then-optional joint, strategic-level Capstone course for promotion to flag officer.
- Improve civilian and military faculty quality.
- Establish a Director of Military Education on the Joint Staff.

The Joint Staff implemented a number of the Panel's recommendations and published the then- titled Program for Joint Education, now the Officer PME Policy (OPMEP).[27] This CJCS policy, which is continually revised, does several things:

- Lays out a general educational philosophy.
- Defines and describes the PME continuum (levels of education or the Skelton "framework").
- Specifies the learning objectives for each level of PME.
- Establishes broad educational standards for all PME institutions.
- Establishes specific standards for military faculty and student body mixes, student-to-faculty ratios, and military faculty quality.
- Establishes a regular and rigorous accreditation process (the Program for the Assessment of Joint Education) for ensuring intermediate- and senior-level PME schools are meeting OPMEP requirements.
- Establishes the Director JCS/DJ-7 as the Deputy Director of the Joint Staff for Military Education (dual-hatted).

According to the Program for Joint Education, 10-month service intermediate- and senior- level education awarded JPME I credit. The then 12-week Armed Forces Staff College (now the 10- week Joint and Combined Warfare School course at Joint Forces Staff

College) awarded JPME II credit. The two senior joint schools (ICAF and National) awarded both JPME I and II credit in "single-phase PJE." Now, almost all of these schools are also accredited to award master's degrees to, at least, U.S. students. Since the 1989 Skelton Panel, the PME system has continued to expand and has become more complex with new institutions and schools. In particular, NDU and Air University have added levels of management, levels of education, and additional education and training programs. Marine Corps University (MCU) has also grown much larger. The CJCS also mandated an increasing emphasis on non-resident programs and created a JPME II-like course for reserve component officers (Advanced Joint Professional Military Education).[28]

In 2005, the Department sought, and the Congress granted, authority for the service senior colleges to award JPME II credit. The Department's rationale was that the services needed more "JPME II complete" officers to create a larger quality pool from which to select general and flag officers, particularly for senior joint duty. In other words, the Department sought to increase throughput. One other change is particularly relevant to this study. Under the direction of the CJCS General Richard Myers, JFSC recently rearranged its school structure. As a result, JFSC now has two schools that award JPME II credit: the 10-week Joint and Combined Warfare School established in 2005; and the 10-month senior-level Joint Advanced Warfighting School established in 2004, which offers credit for both JPME I and II and a master's degree.[29]

The Skelton Panel's recommendations, which have largely been acted upon,[30] were made shortly before the fall of the Berlin Wall in November 1989 and more than a decade before the events of September 11, 2001. Today the military is half the size it was in 1989. The military has fought in two wars for nine years. Arguably, it costs more to achieve less and the joint environment has changed. Acknowledging the variety of national security challenges that have emerged in the intervening years since the Skelton Panel, the Subcommittee examined the officer in-residence PME system in light of the demands posed by a dynamic security environment, and to evaluate whether the system is educating agile and adaptable leaders and thinkers who can meet these demands.

The Security Environment

> *"There is, first of all, a compelling need to develop new ways of creating military advantage in the face of current geopolitical and technological trends.*
> *... In recent years, whether it be 9/11, Afghanistan or Iraq, we have found ourselves reacting to emerging challenges rather than anticipating them. Ignoring growing challenges to our security will not make those challenges go away. Sooner or later, they will have to be confronted."* [31]
>
> Dr. Andrew F. Krepinevich

PME exists to prepare officers to perform effectively within a changing security environment. While certain elements of PME will remain timeless, others must continuously adapt to evolving strategic conditions. In recent years, the security environment has been characterized in government documents as one of increasing complexity, uncertainty, and ambiguity. It is fraught with foreseeable and wide-ranging strategic challenges, yet it presents unprecedented strategic opportunities.[32] Strategic challenges include threats posed by: "violent extremist movements, the spread of weapons of mass destruction, rising powers with sophisticated weapons, failed or failing states, and increasing encroachment across the global commons (air, sea, space, cyberspace)."[33] Strategic opportunities may include international capacity and partnership building, cooperative management of the global commons, concerted nonproliferation efforts, and equitable resource management. All facets of the emerging security environment will place demands on military officers that will need to be addressed through professional education, training, and development.

The Skelton Panel cited combat effectiveness as the principal reason for PME. It noted: "The panel believes that the major subject of professional military education should be the employment of combat forces, the conduct of war. Other subjects such as leadership, management, and executive fitness are useful, but should be secondary."[34] Arguably, the task of producing military effectiveness within the officer corps has expanded and become more demanding since the Skelton Report was published. Because the security environment has evolved considerably since 1989 and promises to continue evolving at increasingly accelerated rates,[35] future strategic opportunities may need to be realized through the performance of roles outside of the military's traditional conduct of combat and combat-related operations. In fact, the Department now stresses that officers "must be *strategically minded, critical thinkers, and skilled joint warfighters*," wherein "the term 'warfighter' is not limited to officers serving in the combat arms."[36] Rather, the term denotes any individual who possesses "'capabilities specific to joint operations' whether in the conduct of war or operations other than war."[37] Military leaders need to possess the intellectual capacities, the mental agility, and the military expertise to operate with diverse partners across a broad range of operations, including those reflective of conventional, irregular, and hybrid warfare, within a continuum embracing pre- and post-conflict considerations.[38]

Capstone Concept for Joint Operations

As a consequence of the demand for the military to perform diverse functions with agility, adaptation to challenges within the evolving security environment has become an enabling principle for educating the men and women who are part of the future joint force. In recent years, the CJCS has provided broad guidance on force development and experimentation that responds to the perceived spectrum of security challenges through the *Capstone Concept for Joint Operations* (CCJO). According to the most recent CCJO:

> The foreseeable future promises to be an era of persistent conflict – a period of protracted confrontation among states, nonstate entities, and individual actors increasingly willing to use violence to achieve their political ends. The future is unlikely to unfold as steady state peace punctuated by distinct surges of intense conflict. Rather, the major initiatives of U.S. foreign policy – major war, strategic deterrence, foreign humanitarian assistance, security cooperation, and so on – are all likely to unfold against a global backdrop of chronic conflict. Such protracted struggles will not lend themselves to decisive military victory, but often at best will be amenable to being managed continuously over time. Many of these conflicts may cut across national, regional, cultural, and combatant command boundaries, complicating the responses to them.[39]

The CCJO not only cautions that the future security environment will require perpetual conflict management among countless pressures, such as religious and ethnic passions, dysfunctional borders, societal collapses, corruption, and natural resource scarcity; it further warns that the threats within the changing security environment are likely to continue growing more pervasive, more diverse, and increasingly dangerous.

> At the same time, the means of waging conflict are becoming more lethal, ubiquitous, and easy to employ. Advanced weaponry, once the monopoly of industrialized states – including anti-access and area-denial capabilities – increasingly is becoming available to both less-developed states and nonstate [actors]. The potential proliferation among a growing roster of states and nonstate actors of weapons of mass destruction, especially nuclear weapons, is particularly dangerous, and could significantly complicate any future U.S. use of military force.[40]

The CCJO further asserts that an effective force posture, sufficiently responsive to the challenges posed by the security environment, will require a preparedness to regularly execute military activities well beyond the conduct of war by test of combat alone.[41]

The CCJO advises that the reality of the complex security environment will require general- purpose forces to perform distinct military tasks apart from, and in addition to, their primary combat roles. Defeating armed enemies, whether regular or irregular, in combat only represents the first of the CCJO's four categories of joint military activity. The joint force will also be required to conduct security, engagement, and relief and reconstruction efforts.[42] Security activities, unlike combat, "seek ultimately to reassure rather than compel."[43] Engagement activities provide cooperative security, but they may have entirely diplomatic or economic contexts of widely varying scope and duration and are subject and sensitive to national and international law, regulation, and standards of comity.[44] Relief and reconstruction activities responsive to events such as combat, civil disorder, or natural disaster are military activities akin to, but distinguishable from, stabilization and reconstruction operations, which are coordinated operations led by the State Department.[45]

According to the CCJO, these four types of military activity encompass "virtually every mission the joint force will be called upon to accomplish" in the security environment.[46] Particulars will vary with context, but the CCJO summarizes its discussion of these military activities by stating:

> Combat, security, engagement, and relief and reconstruction must all be competencies of the joint force. While some special-purpose forces will specialize in particular aspects of one or more, general-purpose forces must be able to operate in all four types of activity in one way or another. Currently, U.S. joint forces possess codified doctrine for the conduct of combat, but doctrine and capabilities with respect to the other activities are less robust. That imbalance must change.[47]

These signals of a doctrinal movement to embrace new, expanded, and rapidly interchangeable roles for general-purpose military forces clearly endorse a broader association of the term "joint" than the Skelton Panel contemplated more than 20 years ago.

Informed by Goldwater-Nichols, the Skelton Report considered joint activities to comprise coordination among the services and integration of their capabilities. However, as described in the CCJO, each of the combat, security, engagement, and relief and reconstruction activities rely heavily on close cooperation and coordination with those of international organizations, coalitions, foreign governments, federal agencies, and state and local authorities.[48] These activities support significant military contributions to joint, interagency, intergovernmental, and multinational (JIIM) operations.

An additional implication for joint force development encapsulated in the CCJO is the need for the joint force to "markedly increase language and cultural capabilities and capacities."[49]

> The idea of understanding each operational situation in its unique political and strategic context will require a higher level of cultural attunement than joint forces currently possess. Similarly, increased emphasis on security, engagement, and relief and reconstruction activities implies even more extensive contact and interaction with indigenous agencies and populations than does combat.[50]

The task of realizing this idea and its associated role requirements is, and will continue to be, heavily reliant on the educational aspect of officer development.

The Joint Operating Environment

The CCJO describes *The Joint Operating Environment* (JOE), periodically issued by U.S. Joint Forces Command (JFCOM) as a companion to the CCJO, which offers detailed observations with respect to the shifting security environment.[51] The JOE divides its perspective into three areas of analysis: influential security-related trends, the contextual bases for potential conflict that are supported by these trends, and the implications for the joint force over the next 25 years.[52] Like the CCJO, the JOE describes challenges that joint forces will potentially face in the future, but many of the conditions it describes exist now. The most recent *National Defense Strategy* concedes: "An underlying assumption in our understanding of the strategic environment is that the predominant near-term challenges to the United States will come from state and non-state actors using irregular and catastrophic capabilities."[53] The JOE considers geopolitical trends in demographics, globalization, comparative economics, energy supply and demand, food production and distribution, water scarcity, climate change and natural disasters, pandemics, cyber connectivity, and the utility of space.[54] It relates these trends to their potential contexts for conflict, including: competition and cooperation among conventional powers (i.e., nation states); challenges and threats to the United States from near-peer competitors (i.e., Russia and China) and regional influences (e.g., NATO, Iran, India, Pakistan); weak and failing states; threats of unconventional power (e.g., non-state and trans- national actors); radical ideologies; the proliferation of weapons of mass destruction; technological advances; rampant urbanization; and, opposing narratives (i.e., strategic communications).[55] Finally, the JOE discusses the implications of these trends and their possible contextual effects on preparing for war, conducting war and other military operations (e.g., activities supporting intelligence, surveillance, and reconnaissance; or deterrence), and, aptly, the opportunities for future growth offered through PME.[56]

Both the CCJO and the JOE recognize PME as a fundamental mechanism for cultivating future military effectiveness. General James Mattis, the current JFCOM commander, stated: "We need an educated, adaptable officer corps, not married to any single preclusive view of war."[57] He also noted that: "we will have to educate better and reward learning in our officer corps, so our leaders can adapt more swiftly than our enemies."[58] This priority is echoed in the JOE, which identifies PME as "the critical key to the future."[59] It suggests that: "All military leaders must be equipped with the confidence to decide and act in ambiguous situations and under conditions where clear direction from above may be lacking or overcome by changing conditions."[60] The CCJO reinforces the call for officers (from the lowest echelons to the highest) with the same attributes, adding, "The Services must recruit, develop, and reward leaders who acquire and demonstrate these skills. Leader development, professional military education in particular, must specifically provide training and education that facilitates flexible and creative problem solving."[61] As to the educational requirement, the JOE further asserts: "This is the fundamental challenge the U.S. military will confront: providing the education so that future leaders can understand the political, strategic, historical, and cultural framework for a more complex world, as well as possess a thorough grounding in the nature of war, past, present, and future."[62]

PME and Officer Development Policy

> *"Service Leader competencies will vary by Service but they are developed in a joint context and are the foundation for joint officer development."* [63]
>
> The CJCS Vision for Joint Officer Development

PME is a fundamental component of an officer's development as a leader and to the development of the armed forces overall. PME's principal purpose is to educate and prepare military leaders, throughout their careers, for the rigorous intellectual demands of employing military forces or other instruments of national power in a complex and uncertain security environment. Military officers at every grade must lead others to perform strenuous tasks in difficult and often dangerous situations. To meet that challenge, the Department and the services created officer development systems, with PME at their core, that endeavor to produce skilled warfighters, who are "strategically minded, critical thinkers."[64] The officer corps must possess the needed competencies specific to the services' primary warfare domains (i.e., the air, land, sea, and space aspects of warfare). The Department and the services must also produce sufficient numbers of officers who can contribute to joint, international, intergovernmental, and multinational operations.

The CJCS and the Service Chiefs have overlapping and complementary responsibilities and authorities with respect to the PME system and officer development. The services bear the legal responsibility of organizing, training, and equipping their respective forces.[65] At the same time, the CJCS serves as the principal military advisor to the Secretary of Defense on JPME matters and is responsible for "formulating policies for the joint training of the armed forces" and for "formulating policies for coordinating the military education and training of members of the Armed Forces."[66] Moreover, the CJCS promulgated a "Vision for Joint Officer Development" (the CJCS Vision for JOD), in which PME is a central pillar of the "joint learning continuum," which includes: education, training, experience, and self-development.[67] The CJCS Vision for JOD is to produce "the largest possible body of fully qualified and inherently joint officers suitable for joint command and staff responsibilities."[68] Ultimately, however, the services control officer management. Each has its own needs for highly qualified officers to serve as commanders within that service and on service-specific headquarters staffs. The services exercise this authority, because they "recruit, commission, educate, and train junior officers in various occupational specialties, and assign, promote, and manage their development from junior to senior officers."[69] The CJCS does not possess the same authority, and, therefore, the CJCS relies on the services to educate officers as joint officers as well as capable service officers.[70] The CJCS Vision for the JOD acknowledges this relationship and establishes, as a key principle, that "*joint officers are built on Service officers*."[71]

While each service is different, their officer development models largely parallel the CJCS model. The services recognize the value of an education that is distinct from, but complementary to, training, and they place special emphases on the importance of PME. Self-development, experience, and mentorship typically comprise the other key components of the services' learning continua for professional development.[72]

The services are responsible for developing officers "with expertise and knowledge appropriate to their grade, branch, and occupational specialty"[73] who can demonstrate the competencies "to meet their own Service-specific roles, missions, and capabilities."[74] The services develop competencies in company-grade and junior officers in core service functions, specific weapons systems, and tactical doctrine. This knowledge is subsequently broadened in field and mid- grade officers to support functions at the operational level (e.g., combined arms and joint campaigns for Army majors and lieutenant colonels and battle group or task force operations for Navy lieutenant commanders and commanders). Finally, the military educates to foster capable strategic- level thinking among its senior officers (lieutenant colonels and colonels, Navy commanders and captains, and flag officers). As the services train and educate their officers in service-specific competencies, their PME also includes instruction in joint matters set by the CJCS in the OPMEP. As officers advance in seniority, the emphasis on joint matters within PME progressively expands from providing basic knowledge of the roles and mission of other services and the military command structures, for example, to graduate education that incorporates more sophisticated analysis of the formulation and evaluation of national security strategy and the development of strategic leadership skills.[75]

Although they differ from one another, each service has a process for ensuring officers' performance effectiveness and tying PME content to their leadership development programs. The commander of the Army's Training and Doctrine Command (TRADOC) oversees the Army's efforts and uses a "Common Core" process to ensure that PME supports the goals set for officers in the *Army Leader Development Strategy*.[76] TRADOC also serves as the Army's executive agent for its "Human Capital Enterprise," which makes it responsible for leader development and allows it to "influence, establish, and change policies that directly affect [Army] leader development and professional military education objectives."[77] The Vice Chief of Naval Operations heads the Navy's Advanced Education Review Board, which is tasked with ensuring that the Navy's education strategy is properly resourced and that education policy is integrated across the Navy.[78] The President of MCU serves as the Marine Corps' central PME proponent and uses a Curriculum Review Board to "manage and link leader development content at each level of PME."[79] The Joint Staff has identified the Marines' Curriculum Review Board process as a "best practice" model for other PME institutions to emulate.[80] In the Air Force, the Officer Force Development Panel, the members of which are "seven three-star general officers, a senior statesman, and several advisors," reviews educational policies and is "focused on how to deliberately develop officers for deep and broad leadership roles, especially those in the joint environment."[81] The Officer Force Development Panel reports to the Force Management and Development Council, headed by the Air Force's Vice Chief of Staff.

In addition to providing oversight for PME in the officer development context, the services coordinate their PME and JPME efforts with the CJCS by several means, including the periodic reviews and revisions of PME curricula that are required by law.[82] The OPMEP establishes feedback and updating mechanisms to examine "PME curricula currency, quality, and validity."[83] Feedback mechanisms include: (1) a requirement that each PME institution has a well-defined curricular review program; (2) periodic joint education conferences for the warfighting community; and, (3) feedback provided by the Military Education Coordination Council (MECC) and its subordinate MECC Working Group. The MECC serves as an advisory body to the Director of the Joint Staff. The principal members of the MECC include: the presidents, commandants, and directors of the joint and service universities and colleges; the heads of any other JPME accredited institution; the Joint Staff's Deputy Director for Military Education (DJ-7); and, the Director of the Joint Warfighting Center at JFCOM.[84] Curricular updating mechanisms "involve all levels of the PME system and the using communities (i.e., Services, combatant commands, and DOD agencies)."[85] Specific updates and mechanisms include: (1) periodic revisions of the OPMEP every five years; (2) regular curricular reviews by each service and joint institution; and, (3) joint faculty education conferences. The CJCS annually recommends "special areas of emphasis" for incorporation into JPME curricula. Special areas of emphasis are based on advice from the Office of the Secretary of Defense, the services, the Joint Staff, and the COCOMs.[86] Topics from the most recent CJCS special areas of emphasis list include: building partnership capacity, countering ideological support for terrorism, defense support for civil authorities, net-centric information sharing, strategic communications, irregular warfare, operational contract support, space as a contested environment, and psychological health awareness.[87] The CJCS also conducts periodic assessments of JPME curricula.[88]

The Joint Duty Assignment List

Historically, JPME has factored in preparing officers to lead in the joint operating environment. Goldwater-Nichols created the joint specialty and Joint Staff Officers (JSOs) who were to serve fluidly in both joint and service-specific assignments. These officers, now referred to as Joint Qualified Officers (JQOs), are "particularly trained in, and oriented toward, joint matters."[89]

"Joint matters" are statutorily defined as:

> … matters related to the achievement of unified action by multiple military forces in operations conducted across domains such as land, sea, or air, in space, or in the information environment, including matters relating to -
>
> (A) national military strategy;
>
> (B) strategic planning and contingency planning;
>
> (C) command and control of operations under unified command;
>
> (D) national security planning with other departments and agencies of the United States; and
>
> (E) combined operations with military forces of allied nations.[90]

"Multiple military forces" may include: other departments and agencies, other countries' military forces or agencies, or non-governmental persons or organizations.[91]

The law also requires the Secretary of Defense to establish "policies, procedures, and practices for the effective management" of JQOs.[92] The Department refers to these policies, procedures, and practices collectively as Joint Officer Management.[93] Goldwater-Nichols directed the Secretary of Defense to publish a list of "joint duty assignments," called the Joint Duty Assignment List (JDAL).[94] JDAL assignments are limited to those assignments in which officers gain "significant experience" in joint matters.[95] An "appropriate number" of JDAL positions must be designated "critical" billets, but "only if the duties and responsibilities make it important that the occupant be" a JQO.[96] The Secretary sets the number of assignments and critical assignments on the JDAL.[97] Before the Joint Qualification System (JQS) was established in 2007, the law required the JDAL to include 800 critical billets and to be of sufficient size to ensure that approximately 50 percent of the JDAL positions were filled by JSOs or JSO nominees.[98] Presently, under the JQS, the JDAL must be large enough to accommodate roughly 50 percent of the officers serving in JDAL billets for the grades of O-5 and above to "have the appropriate level of joint qualification."[99]

Upon passage of Goldwater-Nichols in 1986, the Department began the practice of automatically including all billets for officers in the grade of O-4 and above at the COCOM headquarters, the Joint Staff, and Office of the Secretary of Defense on the JDAL.[100] According to the congressionally mandated "*Independent Study of Joint Officer Management and Joint Professional Military Education*" conducted by Booz Allen Hamilton in 2003:

> Admiral William J. Crowe, Jr., then Chairman, JCS, believed it important to get a JDAL established as soon as possible to avoid any appearance of foot-dragging in setting up [Joint Officer Management]. ... Admiral Crowe thought that the size of the first JDAL of slightly more than 8,200 positions could satisfy all arguments. "We knew some positions wouldn't qualify, but we were afraid of setting up an elite that really wasn't justified." He expected that the list would be further refined to identify the operational positions, but service objections to protect career paths and political fears of creating a de facto general staff prevented it.[101]

In 1989, the Skelton Report suggested that the JDAL "should be both improved and reduced significantly" and suggested a "position-by-position review."[102] The Booz Allen Hamilton study highlighted some of the consequences of the automatic inclusion approach and the Department's failure to apply a meaningful "joint matters" test:

> However necessary for organizational reasons, these decisions went beyond the definitions of joint matters in law and the definition of a [joint duty assignment] in DoD policy. They set the precedent of extending to all staff officers, without regard to their specialty or duties, the strategy, planning, [command and control], and integrated employment functions of the Secretary of Defense, Chairman, JCS, and [combatant commanders]. They equated all staff duties with "joint matters" and thus diluted the understanding and purpose of that term.
>
> A few examples illustrate how current practice strays from the definition in Title 10. The current JDAL includes positions for a deputy comptroller, a morale/welfare/ recreation staff officer, an assistant director of advertising, public affairs officers, directors of military equal opportunity policy, budget analysts, cost analysts, directors of military compensation, and other officers in positions far removed from strategy,

planning, integrated employment of forces and command and control. Service in these positions qualifies officers as JSOs and for promotion to general or flag officer.[103]

The JDAL continues to automatically include all O-4 and higher grade officers from the organizations mentioned above, and it now includes all of the Defense Agencies' headquarters.[104] Despite the Skelton Panel's observation that the JDAL should be significantly reduced, the JDAL has increased. There are 11,730 positions on the current JDAL.[105] However, the Joint Staff reports that a JDAL validation board convened by the Principal Deputy Undersecretary of Defense for Personnel and Readiness has begun a five-year process of reviewing the current JDAL to determine whether specific positions should be retained on, removed from, or added to, the JDAL.[106] The validation board is using a two-part validation test. A reviewed position must: (1) reside in a "joint" organization; and (2) entail a preponderance of duties involving significant experience in "joint matters," using the statutory definition cited above.[107]

The COCOMs, collectively, control more than one-half of the joint duty assignments on the current JDAL. Out of 11,730 total JDAL positions, COCOM positions account for 6,695 positions, or 57 percent.[108] The Defense Intelligence Agency has 910 JDAL positions or the rough equivalent of a larger COCOM headquarters staff.[109] The Office of the Secretary of Defense, the Joint Staff, and the Defense Agencies also have sizeable numbers of JDAL billets. Below is a breakdown of the 2008 JDAL by grade, on which the majority of billets are filled by O-4s and O-5s.[110]

2008 JDAL Positions by Grade	
O-4	4858
O-5	4614
O-6	1992
O-7	130
O-8	80
O-9	41
O-10	15
TOTAL	**11730**

The Joint Qualification System

The Department sought "to change significant aspects of joint officer management and joint military professional education enacted as a result" of the Skelton Report.[111] In 2004, the Committee expressed concern over the Department's lack of a "coherent, comprehensive context" and an "overall vision for joint officer management and education," and it directed the Department to draft a strategic plan that would provide "the framework within which to consider, what, if any, future changes to joint officer management and joint professional military education, are required."[112]

The Department submitted its plan, entitled the *Department of Defense Strategic Plan for Joint Officer Management and Joint Professional Military Education* (the *Strategic Plan)*, in 2006. It heavily emphasized the importance of maximizing joint experience within an officer's career, but it did not reinforce or emphasize a connection between JPME and joint duty assignments, and it offered limited guidance as to the specific roles JPME should play in preparing officers for joint duty. It stated: "Nominally, JPME I should be completed prior to promotion to lieutenant colonel or commander; JPME II should be completed prior to promotion to colonel or captain."[113] The *Strategic Plan* did not tie that education to joint duty assignments or joint matters, whether they are reflected on the JDAL or not. According to the *Strategic Plan*, "[j]oint experience accrues where jointness is applied," not through a "static list of joint duty assignment positions."[114]

Congress then amended portions of Goldwater-Nichols, significantly revising many of the personnel provisions contained in Title IV of the Act. Amendments included removal of the requirement that officers seeking the joint specialty complete JPME I and II prior to a joint duty assignment. The National Defense Authorization Act for Fiscal Year 2007 (FY 2007 NDAA) gave the Department the authority to replace the original JSO system with a new system incorporating different levels of "joint qualification" within JQS.[115] The Department began to implement the JQS on October 1, 2007.[116] Title IV of Goldwater-Nichols sought to "establish policies, procedures, and practices for the effective management of officers…who are particularly trained in, and oriented toward, joint matters" as members of "the joint specialty."[117] The JQS seeks to "transition from a system where the Joint Specialty Officer (JSO) designation is the only recognized level of joint capability to one that offers various levels of qualification based on joint experience."[118] One notable policy change allows officers to apply to receive joint credit for experience with joint matters garnered in non-JDAL assignments, such as joint or interagency deployments, in addition to receiving joint credit for completing 24- to 36-month JDAL assignments.[119]

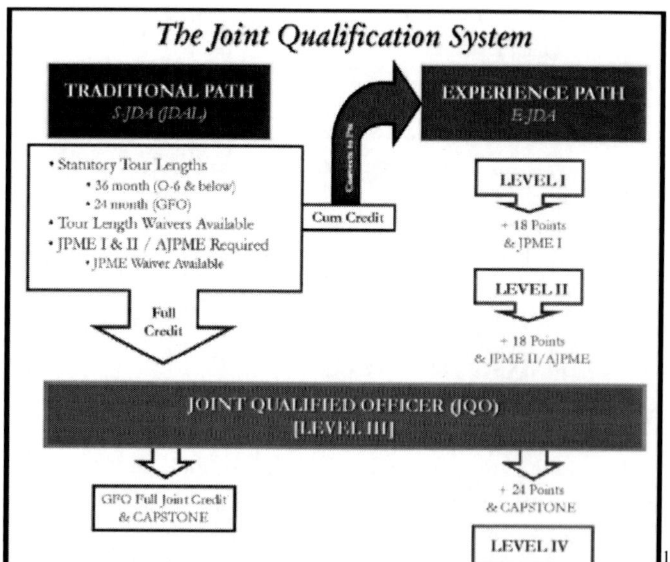

120

The JQS establishes four distinct qualification levels, with each level specifying essential joint experience and joint educational criteria. Title 10 states, "The purpose of establishing such qualification levels is to ensure a systematic, progressive, career-long development of officers in joint matters and to ensure that officers serving as general and flag officers have the requisite experience and education to be highly proficient in joint matters."[121] While standard JDAL assignments are "still the primary means of achieving joint experience and joint duty credit," the JQS recognizes that "attaining expertise in joint matters is a career long accumulation of experiences that may be gained via various duties and assignments or [sic] to joint organizations for extended periods of time or through the performance of temporary duties of shorter duration."[122]

Under the JQS, a commissioned officer achieves Joint Qualification Level I (JQL I) when he or she completes an officer basic course that introduces joint concepts and joint awareness.[123] An officer may then begin accumulating joint qualification points, which may be gained through "joint experiences, joint training, as well as other education determined by the Chairman of the Joint Chiefs of Staff."[124] Achieving Joint Qualification Level II (JQL II) requires: (1) the completion of JPME I; and, (2) the completion of a JDAL assignment or accumulation of 18 joint qualification points.

The term "JQO" is synonymous with Joint Qualification Level III, and it supersedes the previous JSO designation. Officers must complete JPME I and II prior to becoming JQOs, but unlike JSOs, they are not required to do so prior to serving in a joint duty assignment. Joint Qualification Level IV, is attained by flag officers who have completed the Capstone course (JPME III) and either completed a joint-duty assignment or accumulated at least 24 joint qualification points.[125]

The Department is in the process of incorporating the new joint qualification levels required for JDAL positions. The Department indicated that a "preliminary assessment" suggests that the majority of JDAL billets will be filled by officers at either JQL I or JQL II. Consequently, officers at JQL I will not be required to complete any JPME prior to serving in a joint duty assignment, while officers at JQL II will only be required to complete JPME I.[126] The amendments to Goldwater- Nichols contained in the FY 2007 NDAA eliminated the requirement for the Secretary of Defense to designate 800 critical JDAL positions. In its place, the legislation granted the Secretary the discretion to "designate an appropriate number of joint duty assignment positions as critical."[127] The Department continues to support roughly 800 critical billets on the JDAL. Those billets will be filled with JQOs.[128]

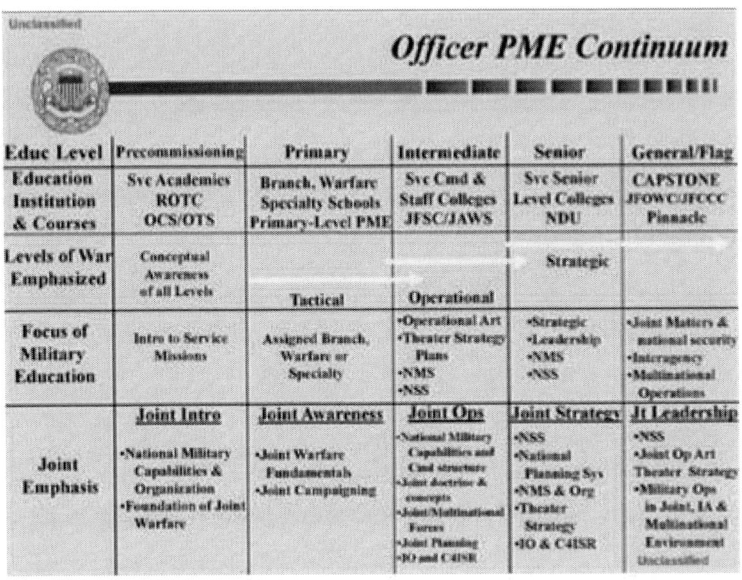
Unclassified

Officer PME Continuum

Educ Level	Precommissioning	Primary	Intermediate	Senior	General/Flag
Education Institution & Courses	Svc Academies ROTC OCS/OTS	Branch, Warfare Specialty Schools Primary-Level PME	Svc Cmd & Staff Colleges JFSC/JAWS	Svc Senior Level Colleges NDU	CAPSTONE JFOWC/JFCCC Pinnacle
Levels of War Emphasized	Conceptual Awareness of all Levels	Tactical	Operational	Strategic	
Focus of Military Education	Intro to Service Missions	Assigned Branch, Warfare or Specialty	•Operational Art •Theater Strategy Plans •NMS •NSS	•Strategic •Leadership •NMS •NSS	•Joint Matters & national security •Interagency •Multinational Operations
Joint Emphasis	**Joint Intro** •National Military Capabilities & Organization •Foundation of Joint Warfare	**Joint Awareness** •Joint Warfare Fundamentals •Joint Campaigning	**Joint Ops** •National Military Capabilities and Cmd structure •Joint doctrine & concepts •Joint/Multinational Forces •Joint Planning •IO and C4ISR	**Joint Strategy** •NSS •National Planning Sys •NMS & Org •Theater Strategy •IO & C4ISR	**Jt Leadership** •NSS •Joint Op Art Theater Strategy •Military Ops in Joint, IA & Multinational Environment

Unclassified

SYSTEM ISSUES

"Almost 80 percent of today's U.S. military officers were accessed after the [Goldwater-Nichols Act] was implemented. One could argue they have 'grown-up' in the joint environment. This first generation of jointly immersed officers are coming of age and rising to the senior ranks. Make no mistake; this did not occur naturally or randomly. DOD, in concert with CJCS, has focused on 'continuously improving joint readiness by aligning joint education and training capabilities and resources with combatant command needs... .'"[129]

Department of Defense Strategic Plan for Joint Officer Management and Joint Professional Military Education

This section of the report looks at significant issues within the larger PME system. These issues center on whether the system is achieving its objectives. The section begins by examining the diminished relationship between JPME and joint duty assignments and the effectiveness of PME as preparation for staff duty assignments. The report also describes tensions that the Subcommittee found within the PME system, most notably, among the competing demands in officers' careers. The report also evaluates current efforts to develop strategists within the officer corps. Finally, the report looks at the delivery of PME curricula in terms of pedagogical practices and rigor, and it examines PME content in relation to how PME curricula balance emerging and enduring subject matter.

The Diminished Relationship between JPME and Joint Duty Assignments

The 1988 House Armed Services Committee Panel on Professional Military Education (Skelton Panel) described "education on joint matters" as "a basic link between a service competent officer and a joint competent officer."[130] The Goldwater-Nichols Department of Defense Reorganization Act of 1986 (Goldwater-Nichols) required the establishment of joint officer management policies and addressed educational requirements for both service and joint officer competencies to strengthen contributions to joint cooperation and prepare officers for duty in the joint arena.[131] Over the years, Congress amended Goldwater-Nichols numerous times to give the Department greater flexibility in conducting joint officer management. The policies, procedures, and practices that have been implemented as a result of this increase in Departmental discretion and expansion of the joint duty assignment list (JDAL) have contributed to the estrangement of joint professional military education (JPME) from its statutory purpose of preparing officers for joint duty assignments.

Joint Education for Joint Specialty Officer Assignments

Goldwater-Nichols connected a joint specialty officer's (JSO's) joint duty assignments to his or her completion of joint education, JPME I and II, in several ways. Goldwater-Nichols established a general sequencing requirement in which JPME and joint duty assignments were to be completed for those officers who sought to become JSOs.[132] The Skelton Report described this sequence as follows: "Officers first go to joint education; they then serve in a joint assignment as a JSO nominee. After successfully completing a full joint tour, they can then be selected as a JSO."[133] At the time, a joint education for JSOs was understood to consist of JPME I and II.[134]

Goldwater-Nichols also established "post-education duty assignments."[135] Unless waived by the Secretary of Defense, the law required every JSO graduating from a JPME school to be immediately assigned to joint duty for his or her next assignment, and it required greater than 50 percent of all other officers (non-JSOs who became eligible to be JSO nominees by virtue of completing the educational program at a JPME school) graduating from those schools to be similarly assigned ("the 50 percent plus one" requirement).[136] Congress later amended the law to allow the Department greater flexibility in meeting that requirement. Non-JSOs were permitted to complete joint duty either in their first or second post-education assignment if the Secretary determined it necessary for "efficient management of officer personnel."[137] The "50 percent plus one" requirement was not extended to the senior service schools when Congress granted them the authority to award JPME II credit in 2005. Consequently, the requirement only applies to those schools within the National Defense University (NDU).[138] The legislation did not apply the requirement beyond the NDU schools because the services sought the flexibility "to fill all joint and internal billets, particularly those in warfighting specialties, with appropriately qualified officers."[139] The "50 percent plus one" requirement, however, continues to provide a link between joint education and joint duty assignments for many NDU graduates.

Joint Education for All Officers

The Skelton Panel also described the challenge of providing "education in joint matters for *all* students, whether or not those students will become JSOs."[140] The Skelton Report explained:

> Establishment of the joint specialty to support the Chairman, [Joint Chiefs of Staff], and the unified and specified commanders does not obviate the need for improving joint education in service schools for officers throughout the armed forces. Even with the emergence of the joint specialist, joint staffs will continue to be manned primarily by non-joint specialists (including inexperienced nominees for the joint specialty). Consequently, non-JSOs need training in joint staff procedures and systems, and broad education in the capabilities, limitations, and doctrines of the other services. *In fact, non-JSOs are essential to the proper functioning of the joint system because they bring current service expertise and credibility to bear in considering the solutions to joint problems.* The Chairman, JCS and the unified and specified commanders – and the joint specialists – will rely upon service experts to elaborate force options and to implement decisions.[141]

The Skelton Panel envisioned the intermediate-level PME schools as "the principal schools for learning jointness."[142] The Skelton Panel suggested: "Everyone who attends service intermediate schools should learn the mechanics of joint matters that all officers should know: other service capabilities, limitations and doctrines, and the relevant joint procedures and processes."[143]

Congress subsequently legislated requirements for PME curricula at the services' intermediate- and senior-level schools and the curricula specifically associated with JPME II to have structured foci on preparing officers for joint duty assignments.[144] Goldwater-Nichols required periodic review and revision of the services' intermediate- and senior-level schools' curricula "to strengthen the focus on – (1) joint matters; and (2) preparing officers for joint duty assignments."[145] Congress also included the policy in the FY 1990 and 1991 NDAA, which described the way in which JPME II should focus on joint matters:

> The curriculum should emphasize multiple "hands on" exercises and must adequately prepare students to perform effectively from the outset in what will probably be their first exposure to a totally new environment, an assignment to a joint, multiservice organization. Phase II instruction should be structured so that students progress from a basic knowledge of joint matters learned in Phase I to the level of expertise necessary for successful performance in the joint arena.[146]

Similar language was codified in the FY 2005 NDAA: "The committee believes these provisions [among others] have a permanence and continuing importance that warrant codification."[147] The law states, in part:

> (b) Phase II Requirements – The Secretary shall require that the curriculum for Phase II joint professional military education at any school –
> (1) focus on developing joint operational expertise and perspectives and honing joint warfighting skills; and
> (2) be structured –
> (A) so as to adequately prepare students to perform effectively in an assignment to a joint, multiservice organization; and
> (B) so that students progress from a basic knowledge of joint matters learned in Phase I instruction to the level of expertise necessary for successful performance in the joint arena.[148]

Observation: An express purpose of intermediate- and senior-level PME is the preparation of officers for joint duty assignments.

Secretary of Defense Robert Gates takes questions at Air War College.

Congress changed the law significantly when it replaced the JSO system with the Joint Qualification System, which governs the process of producing joint qualified officers (JQOs).[149] In 2006, Congress removed the requirement for officers to complete JPME I and II prior to serving in joint duty assignments while in the process of becoming a JQO.[150] By granting the Secretary of Defense the discretion to prescribe and administer the various joint qualification levels within the Joint Qualification System, Congress gave the Department the flexibility to conduct joint officer management to a large degree through departmental policy and oversight.[151] However, Congress did not alter the statutory requirement that JPME provide a mechanism for preparing officers for joint duty assignments.

The Department has exercised its discretion in a manner that has distanced JPME, and, in particular, JPME II, from joint duty assignments. This is a striking development, considering the central role JPME previously played in joint officer development. In 1989, the Skelton Panel estimated that the JDAL consisted of 8,300 positions.[152] Prior to implementation of the Joint Qualification System, completion of JPME had distinguished JSOs and JSO nominees from other officers. Congress considered JPME II to be essential preparation for effective performance in a significant number of joint duty assignments, since, with certain limited exceptions, one-half of the JDAL billets had to be filled with officers having completed JPME I and II as JSOs or JSO nominees.[153] Rather than reducing the JDAL, as the Skelton Panel recommended, the Department has expanded it by over 41 percent to approximately 11,730 positions.[154] This expansion is reflective of continual growth in the overall joint force, despite comparative reductions in the services' individual and collective personnel end strengths from those of twenty years ago. Under the Joint Qualification System, the requisite joint education and joint experience requirements for the various joint qualification levels are applied to this broader joint community via departmental policy, while the law requires that approximately 50 percent of the officers serving in JDAL billets designated for O-5s and above "have the appropriate level of joint qualification."[155]

As a result of these changes, it is no longer clear what relationship JPME has with joint duty assignments. Although the Joint Qualification System extends its requirements (including JPME) to a broader officer population than the joint officer management system that preceded it, the law, as amended since Goldwater-Nichols, corresponding departmental policy, and more extensive JDAL demands have combined to allow the Department to satisfy those greater JDAL demands with lesser qualified personnel. The law requires PME and JPME curricula to prepare officers for joint duty assignments, but outside of the limited context of critical billets, it no longer requires officers to attend JPME prior to their joint duty assignments. The Department indicated that it is reevaluating the correlation between joint qualification levels and JDAL assignments, and the Joint Staff offered a preliminary assessment. The Joint Staff indicated that the services will continue to develop officers to the educational and experiential specifications of the differing joint qualification levels, but it expects most JDAL billets will not require successful completion of JPME II. The Joint Staff reported that over a five-year period through September 2008, 1,041 officers completed JPME II following their JDAL tour, and it specifically noted that JPME II is not "a prerequisite for a JDAL assignment."[156] The Joint Staff also indicated that many JDAL billets will not require completion of either JPME I or JPME II. Under the Joint Qualification System, it is likely that only critical JDAL billets will require officers to have completed JPME I and II.[157] As a consequence, departmental policies, procedures, and practices do not appear to consider JPME as vital preparation for all joint duty assignments, notwithstanding the purposes of PME and JPME expressed in law.

Moreover, another significant change in law has also helped to compound the situation and to further displace the link between JPME and joint duty assignments. In 2001, Congress amended Goldwater-Nichols to require officers to be designated as JSOs, and now JQOs, before they can be appointed to become a flag officer.[158] Congress gave the services several years to prepare for this change, and the amendment ultimately became effective in October 2008.[159] This change, combined with those changes eliminating the requirement for completion of JPME I and II prior to serving in a joint duty assignment, support the observation that JPME is regarded as more relevant to promotion than to preparing officers to perform effectively in joint duty assignments. The marked disassociation between JPME and joint duty assignments should be examined closely because it calls the very purpose of JPME into question.

***Finding*: Due to changes in law and policy, JPME, and especially JPME II, appears to be more relevant to enabling officers to compete for promotion into the flag officer ranks than for preparing officers for joint duty assignments.**

***Finding*: Former connections supporting JPME I and II as preparation for joint duty assignments have been substantially weakened. Aside from the limited requirement to fill critical JDAL billets with JQOs, law and policy do not require officers to receive JPME prior to serving in a joint duty assignment.**

***Recommendation*: The Secretary of Defense, the Chairman of the Joint Chiefs of Staff (CJCS), and the Service Chiefs must either implement policies, procedures, and practices for reinforcing the relationship between JPME and preparation for joint duty assignments or show justifiable cause as to why they cannot. In doing so, they should evaluate how a sequential linkage between prerequisite JPME (at each successive phase) and appropriately corresponding joint duty assignments could be established. They should also evaluate how JPME content, and especially JPME II content, should be structured to better fulfill its statutory purpose as preparation for effective performance in joint duty assignments. The Secretary of Defense should report to Congress on the findings and recommendations of this departmental effort.**

PME as Preparation for Staff Duty Assignments

PME has been evaluated with respect to preparing officers for joint and service-specific staff duty assignments. Joint staff officers may not be receiving sufficient PME. Furthermore, PME may have a less than desired qualitative impact on joint and service-specific staff officer proficiencies.

During the course of the Subcommittee's review, various Combatant Command (COCOM) headquarters' staffs drew attention to a study that the Joint Staff commissioned in 2006 to "determine the competencies joint staff officers need for successful job performance."[160] This study was published in April 2008, and it is often referred to as the "Fenty Study" after its author Dr. Linda Fenty. The Joint Staff initiated the Fenty Study "due to continuing requests from the Combatant Commands for targeted training to properly prepare officers to work at the proficiency levels needed within an executive level joint environment."[161] The Fenty Study reported: "Consistently, leaders remarked that most staff officers are arriving at Combatant Command Headquarters (HQ) without the knowledge, skills and abilities needed to perform their tasks in a strategically focused work environment."[162] The Fenty Study further observed: "After analyzing the data, it is clear that targeted training and *education* for becoming a successful staff officer appears to be virtually non-existent – or fragmented at best – prior to arriving at a Combatant Command headquarters."[163]

Timing and Content

The legal requirement for JSOs to complete JPME I and II prior to a joint duty assignment was eliminated in 2006 when Congress authorized the Department to create the Joint Qualification System.[164] More than 80 percent of JDAL billets are designated for officers in the grades of O-4 and O-5.[165] The Joint Staff reported that departmental policy and practice support the filling of billets with officers who are one grade below the designated grade. As a result, significant numbers of JDAL billets can be filled by O-3s and junior O-4s.[166] For example, approximately 49 percent of the joint staffing billets at the COCOMs are authorized at the grades of O-4 and below.[167] This practice may be due to the services having to meet increased joint and service-specific personnel requirements in a prolonged era of high operational tempo. Amid these pressures, the services often opt to fill billets intended for senior officers with junior officers. These junior officers may have only completed pre-commissioning- and primary-level education. Officers in the grade of O-3, typically, would not have attended intermediate-level education (ILE), which includes JPME I. Depending on their seniority, the O-4s in question could have completed JPME I and possibly JPME II, but they are unlikely to have completed both and may not have completed either. With regard to JPME, the Fenty Study made the following observation:

> Staff officers and leadership, like their peers in the other commands, are concerned about the Joint Professional Military Education (JPME) process. Participants did not differentiate education from training when they discussed ***what*** needed to be learned, but almost all were of the opinion that JPME should be the critical venue for providing staff officers with the appropriate joint knowledge with a solid foundation in strategic thinking and writing skills. Almost all participants believe staff officers should attend JPME before arriving at a Combatant Command, instead of after.
>
> Currently, according to the survey participants, officers feel the system is broken because it is the exception instead of the rule that a staff officer gets to attend JPME prior to a Combatant Command assignment.[168]

The Fenty Study also characterized staff officers and senior leaders as "adamant" that the service-specific aspects of ILE, at a minimum, are needed "since one of the areas of expertise a staff officer is supposed to bring to the job is knowledge of his or her Service. Without the ILE program experience a staff officer lacks some of the deeper understanding of the capabilities, tools, processes, and culture of his or her respective Service."[169] However, 75 percent of the O-4s participating in the study, and, presumably, an even higher percentage of officers above the grade of O-4 had completed ILE, which begs the question as to whether ILE (with JPME I) is sufficient preparation for joint duty assignments at the COCOMs.[170]

***Finding*: Amid increasing joint and service-specific staff duty requirements, significant numbers of officers serving in JDAL billets are too junior to have attended ILE (with JPME I) prior to serving in their joint duty assignments.**

The services also send officers to joint duty assignments who, while senior enough to have attended JPME II, have not completed the course.[171] In what appears to be a tug of war with the services, COCOMs are increasingly reluctant to leave a position vacant while an officer attends JPME II for 10 weeks. Moreover, they are increasingly unwilling to allow officers to attend the 10- week JPME II course while they are assigned to those COCOMs. They contend that these officers should arrive at their assignments having completed JPME II.

Officers who had attended JPME II prior to a joint duty assignment gave some indication that the course did not prepare them adequately for those assignments. The relatively few officers who had attended JPME II gave it a mixed review: "Assessments from those who had attended JPME II were equally divided; one third said the course was extremely helpful, one third said that some parts were helpful, and one third said it was little or no help at all."[172] On a scale of zero to five, with zero being "No Help" and five being "Exceptional," officers who had attended JPME II at the 10- or 12-week course or at the war college-level 10-month courses, ranked its helpfulness at 2.7 and 2.8, respectively.[173] Interestingly, ILE with JPME I ranked higher in usefulness at 3.2, possibly due to its curricular focus on planning.[174]

The Fenty Study did not identify the number of junior officers who served in billets designated for more senior officers and who had not received ILE or JPME I. The Study suggested, however, that this situation was probable in a sizeable number of positions.[175]

***Finding*: One-third of officers surveyed by the Fenty Study who had completed JPME II prior to their joint duty assignment considered the course to be without preparatory value.**

***Finding*: Officers surveyed by the Fenty Study who had completed JPME II prior to their joint duty assignment at a 10- or 12-week course rated the usefulness of their education as preparation for joint duty assignments on a par with that rated by officers who had completed JPME II at a 10-month course.**

***Finding*: Officers surveyed by the Fenty Study who had completed ILE (with JPME I) prior to their joint duty assignments rated JPME I's usefulness higher than officers surveyed who had completed ILE (with JPME I) and JPME II prior to their joint duty assignments rated the usefulness of JPME II.**

As previously established, there is no legal requirement that JPME be completed prior to assignment as a joint staff officer, unless the billet in question is designated as a "critical" JDAL billet.[176] In the past, if an officer was pursuing the joint specialty, the standard sequence was to complete JPME I and II and then a joint duty assignment.[177] Although the law no longer requires JPME prior to a "non-critical" joint duty assignment, it does require that JPME II curricula be structured "to adequately prepare students to perform effectively in an assignment to a joint, multiservice organization."[178]

The law also requires the Secretary of Defense to ensure that the intermediate- and senior- level service schools periodically review and revise their curricula, which includes JPME I and II respectively, "to strengthen the focus on (1) joint matters; and (2) preparing officers for joint duty assignments."[179] The Subcommittee heard repeatedly that JPME is not intended to prepare officers for their job as a staff officer.[180] However, the PME and JPME curricula are intended, in part, to prepare officers for joint duty assignments. Most JDAL positions are joint staff officer positions.[181] It is logical to expect that, given the law, PME and JPME would provide some degree of preparation for joint staff assignments. Remedies to certain shortcomings identified by the Fenty Study may be administered through education.

While the Subcommittee recognizes that the Fenty Study is not necessarily definitive, aspects of it are relevant to an examination of the timing and efficacy of PME. The Subcommittee received no indication that the Joint Staff or the services disagree with the issues raised by the study. The Joint Staff cautioned, however, that the study's significance should not be misinterpreted or overstated as its purpose was to identify needed *training* outcomes as opposed to addressing issues involving PME. The Joint Staff also expressed its concern that the study did not consult educators. Rather, the Joint Staff asserted that it relied on the critiques of joint staff officers, who may not be in the best position to distinguish whether the challenges they faced would be better addressed through training or education.[182] Furthermore, the combatant commanders themselves did not express similar concerns about the preparation of their joint staff officers when they met with Subcommittee members. The PME schools also survey graduates and their supervisors. The Subcommittee did not review these surveys, but their results should be included in the full context of evaluating the effectiveness of PME as preparation for joint duty assignments. The Subcommittee notes that the Fenty Study, at the very least, raises the question of whether PME, joint or otherwise, can play a role in alleviating the issues that it identified.

Thus far, the Joint Staff has responded to the Fenty Study by developing tools and materials to better train joint staff officers and by attempting to standardize elements of staff processes across the COCOMs. For example, the Joint Staff developed the *Joint Officer Handbook, Staffing and Action Guide*, which contains important and helpful reference materials.[183] The efforts made by the Joint Staff in response to the Fenty Report are necessary, but insufficient.

Needed Competencies

The Fenty Study focused on joint staff officer competencies. "One of the most important skills sets for joint staff officers, according to leadership, is the ability to accurately assess a task, research appropriate background information, concisely provide optional courses of actions, make recommendations to senior leaders, and factually support recommendations."[184] While senior leaders expressed their respect for "their staff officers' commitment and energy," they voiced concern that "currently some O-4s and O-5s appear to not understand the basics/fundamentals of staff work, and that even O-6s who have not had prior joint assignments are having difficulties."[185] More specifically, among the particular areas highlighted for improvement, the most relevant to the Department's education system appears to be the need for staff officers to:

> *Develop a Better Understanding of the Role and Work Requirements of a Joint Staff Officer.* They need the ability to think in terms of broader objectives, without always focusing on their own specific areas. They need to be able to develop and foster strong interpersonal relationships with other COCOM counterparts. They need a solid understanding of what questions a staff officer needs to ask, and be capable of responding

to taskers rapidly with an all-encompassing approach. They need to understand that their function is to identify a problem, analyze it, identify [courses of action] and make recommendations suitable for a [general or flag officer].[186]

Surveyed staff officers shared a similar concern. They recognized the "need to analyze and synthesize large amounts of information into a concise, brief format for senior level review."[187]

Another part of the Fenty Study used senior leadership feedback to identify 15 competencies that joint staff officers should possess in order to succeed. The purpose of the list was to provide "a baseline for which education and training solutions can be targeted."[188] Some of the competencies can be achieved through training alone. Others, however, are more suited to being strengthened through education, rather than solely through training. Those competencies include the following:

Competency #1: Understands the role of a joint staff officer, and performs work requirements consistently at a high level of proficiency.[189]

Competency #4: Is highly knowledgeable of his/her Service organization, capabilities, and business practices.[190]

Competency #7: Able to write, read, and conduct research at an advanced level appropriate for work performance at an executive level.[191]

Competency #8: Uses well-developed strategic and higher order critical thinking skills for task assignments and problem solving.[192]

Competency #10: Able to communicate effectively at executive levels and across a diverse workforce.[193]

Competency #14: Able to effectively participate in exercise preparation/planning.[194]

Most of the joint staff officers that participated in the Fenty Study had already completed ILE, which includes JPME I.[195] This would suggest that ILE and JPME I are insufficient preparation for joint duty assignments at the COCOMs. The Fenty Study would also suggest that the content delivered at ILE needs to be modified to better develop staff officer competencies. In addition, because significant numbers of O-3s serve on COCOM staffs, consideration should be given to developing needed competencies at the primary and, perhaps, even the pre-commissioning educational levels. The Subcommittee recognizes that PME at the O-3 level properly focuses on developing needed competencies in an officer's primary military occupational specialty. However, if O-3s are being assigned to joint billets, additional preparation may be necessary.

The deficiencies identified in the Fenty Study are not necessarily exclusive to JPME and joint duty assignments. For example, the Fenty Study indicates that joint staff officers may not possess the appropriate level of critical thinking and strategic writing skills needed to perform their jobs effectively.[196] In the Subcommittee's view, development of these competencies should not be exclusive to JPME. Rather, these abilities should be progressively developed and honed throughout an officer's education (i.e., throughout service-specific PME, joint PME, and undergraduate and graduate studies) and over the course of a career.

Another study deserves mention in this regard. A 2008 study performed by the Center for Naval Analyses (CNA), entitled *Developing and Education Strategy for URL [Unrestricted Line] Officers*, examined the "requirements for assignments on [Navy] staffs of operational commanders."[197] It produced findings on naval officer competencies similar to those of the Fenty Study. For example, the CNA study was undertaken due to the "widely held belief that the Navy does a good job developing officers within their warfare communities but a less effective job of preparing them for the later stages of careers, when assignments require a variety of expertise beyond primary warfare areas."[198] The CNA study identified eight areas of expertise needed for the Navy's operational staff officers, in which "they were deficient."[199] Those areas echoed many of the competencies identified as deficient in the Fenty Study. These areas included "critical thinking," "written and oral communication," "knowledge of other services," "knowledge of joint operations," "broad knowledge of the Navy," "expertise in operational planning," "cultural awareness," and "expertise in fiscal issues."[200] The Subcommittee cites the CNA study because it suggests that the issue of ineffective preparation of staff officers may also apply to service-specific staff officers and to PME across the board.

***Finding*: ILE and JPME I are insufficient preparation for joint duty assignments.**

***Finding*: Although the Joint Staff has initiated a training-based response to the findings of the Fenty Study, many of the competencies highlighted in both the Fenty and CNA studies should be addressed through professional military education.**

***Finding*: PME and JPME may require more emphases on competencies needed by staff officers if they are to contribute to effective staff officer performance in service-specific and joint staff assignments.**

***Recommendation*: The Secretary of Defense, the CJCS, the Service Chiefs, and the Joint Staff should develop remedies for the shortcomings identified by the Fenty and CNA studies that targeted education, training, and modifications to relevant personnel processes. Officers should complete appropriate education before they are assigned to a joint or senior service staff.**

Availability and Throughput

The services are assigning officers to joint billets who are senior enough to have attended JPME II but have not yet done so.[201] Three of the services contend that their policy or priority is to send officers who have completed JPME II to joint duty assignments, but they are constrained by the limited availability of seats at the 10-week JPME II course, which they refer to as a "JPME II throughput issue."[202] The Navy has an alternative view, underscoring that "[t]here are no prerequisites to fill non-critical [Navy Joint] billets."[203] The Navy reports that it makes "every effort" for officers "targeted for joint critical billets" to either be JQOs or to have their officers schedule and complete JPME II prior to reporting. The Navy also works to get officers JPME II en route to the [non-critical joint] billets as well, but it acknowledges that competing demands can prevent JPME II completion prior to joint duty assignments. The director of the Navy's training and education division noted that: "the need for our front-running officers, our future leaders, to maintain tactical and operational proficiency, gain leadership and command experience, and pass war fighting skills to our junior members compete with JPME II and, in limited cases, preclude this education en route."[204] The Navy's position illustrates that competing operational needs may be more pervasive than availability constraints in limiting JPME II throughput for the purposes of preparing officers for joint duty assignments.

However, demand at the COCOMs for JPME II-qualified officers appears to be pronounced. The COCOMs' have registered varied responses to the arrival of officers who have not completed JPME II prior to reporting for duty. A number of COCOMs have issued policies limiting JPME II attendance while assigned, namely, U.S. Southern Command (SOUTHCOM), U.S. Central Command (CENTCOM), and the North American Aerospace Defense Command (NORAD) and U.S. Northern Command (NORTHCOM). For example, NORAD and NORTHCOM issued a policy memorandum asserting that officers should attain JPME II credit either "via successful completion of a Service War College or equivalent" or at the 10-week Joint and Combined Warfare School (JCWS), "en route to a joint assignment."[205] Attending JPME II while serving a joint tour with NORAD and NORTHCOM "should be kept to an absolute minimum."[206] CENTCOM's policy prefers that officers needing JPME II attend en route. Officers who have not attended may attend in their first year of a CENTCOM assignment with the approval of their staff directors. Officers who are beyond their first year of assignment require the approval of CENTCOM's Chief of Staff (i.e., a two-star flag officer).[207] Raising the level at which approval may be granted serves to limit policy exceptions. On the other hand, CENTCOM's Chief of Staff saw the policy as progressive in granting directors the discretion to approve temporary duty to the 10-week course for officers in the first year of their assignment.[208] SOUTHCOM also instituted a policy limiting JPME II attendance during an assignment in SOUTHCOM.[209]

This situation may be further complicated by the requirement for officers to have the JQO designation prior to becoming eligible for appointment to flag officer. Although JPME II credit is widely available at the senior service schools, a talented pool of promotable officers either does not attend a senior war college in-residence or receives senior-level education through a variety of fellowship programs that do not confer JPME II credit. The law requires that JPME II be taught in residence, but the 2005 CJCS Vision for Joint Officer Development stated that hybrid learning techniques, blending resident and non-resident delivery methods, would eventually be implemented "to extend the benefits of JPME [II] to the largest possible number of officers."[210] The Department has reported that it may request that Congress extend it the authority to provide JPME II instruction via these hybrid learning methods through the Joint Forces Staff College.[211] These promotion requirements and competing career demands have reportedly caused a temporary influx of O-6s in need of attending the 10-week JPME II course, which has the potential to displace more junior officers en route to joint duty assignments.[212]

***Finding*: Although the demand for officers who have attained JPME II credit via the 10- week course at Joint Forces Staff College (JFSC) appears to be greater than the throughput that institution can support, it is unclear whether this demand is more closely associated with preparing officers for joint duty assignments or with enabling them to compete for promotion to general or flag rank.**

System Tensions

Competing Demands

There is tension between the officer's assignments necessary for career development, the needs of the joint force, and professional military education, whether it is at a military institution or a civilian institution. The competing demands over the course of a 20- to 30-year career make it difficult under normal conditions to accommodate the need for the requisite education, training, and experience; the prolonged contingency operations has exacerbated the tension. The expansion of the number of joint billets on the JDAL and the tightening of the requirement of joint qualification status in order to be eligible for flag rank are among the factors that have put pressure on an increasing number of officers to complete JPME and to gain the requisite joint experience.

For the most part, the officer career development or progression system for the services is based on their year of commissioning. Officers commissioned in the same year normally progress through assignments and promotions with their peers. There are exceptions for a small few who will be promoted earlier than their peers or later than their peers and each of services may have slightly different lengths of service for each rank, but the systems are similar enough. Most officers are focused on what assignment or specific job is required for advancement to the next rank or position of higher responsibility. PME is a part of this focus if it is seen as a requirement for promotion. Each of the services view attendance at the levels of schools differently and may put a slightly different emphasis on each. For example, the Navy generally prefers to send its officers to either intermediate or senior PME in-residence instead of both, in part because of a service culture that values extensive experience in naval operations in the fleet. The Army by contrast has selection boards for those who will attend senior schools. This was previously the case for intermediate-level education, until the Army changed its system and expanded the opportunity to all officers at the rank of major. One could argue that the majority of officers in all the services perceive that in order to remain competitive for promotion, they should spend as much time as possible in operational, preferably deployed, billets, and subsequently less time on high-level staffs or in school.

A study of the military 10 years after the Skelton Panel described the pressures on officers to fulfill both joint and service requirements:

> An Army officer has approximately twelve years from the time he can begin the process of becoming a JSO until the time that he either makes brigadier general or retires, and these twelve years are filled with the types of assignments likely to win promotion to general such as command, service headquarters staff, and service school rotations. Officers would thus lose one quarter of their within-service assignments in order to fulfill their joint duty rotation and JSOs would lose one third.[213]

The fact that the observation was made a decade ago does not diminish its merits given that the 12- year window has likely shrunk. Not only has this compression occurred but some command tours have increased in length. General Martin Dempsey, commander of the Army's Training and Doctrine Command and recently appointed as the Army's executive agent for the Army's Human Capital Enterprise,[214] noted that tour lengths of up to 40 months for brigade commanders were having ripple effects resulting in squeezing out leader development opportunities. He recommended that they be cut back to two years.[215] These system pressures stand apart from the innate value and necessity of joint competency in an environment described as one in which "(e)very officer is likely to be affected to some degree by joint considerations."[216] This strain has worsened, in many cases, due to a decrease in "dwell time" at home stations between deployments overseas, which reduces the available window for in-residence joint education or even distance learning.

As a result of these demands, the services have sought flexibility through modifications to the PME system since the Skelton Report. One example is the establishment of multiple paths for accruing joint duty credit. Previously, the only means of completing the joint experience requirement was by serving 36 months in a JDAL billet, waivable under certain conditions to 22 months. Officers in newly created headquarters and units in Iraq and Afghanistan found themselves in "temporary" positions dealing with "joint matters," but these positions were not listed on the JDAL and thus did not receive joint credit. The Joint Qualification System now governs the accumulation of joint qualification points for joint experience in non-JDAL billets that entail duties involving joint matters.[217] In fact, in as little as a year, it is possible for officers performing duties in Iraq and Afghanistan to acquire joint qualification points sufficient to be the equivalent of a full three-year tour of duty in a standard JDAL billet.[218]

One area in which the services are seeking additional relief is in the creation of a distance- learning avenue for JPME II credit. Although the Department's strategic plan for PME suggests that delivery approaches that use hybrid or blended techniques may at some time be implemented for JPME II and goes as far to say that "(c)urrent [joint officer management] and [joint officer development] needs cannot be met with existing practices,"[219] it remains to be seen whether a strong enough case can be made for proceeding with this option.

These service efforts to seek flexibility in managing all the requirements of an officer's career are necessary and understandable. The fact remains, there is only so much that can be done in a fixed amount of time – say 20-30 years. In the Subcommittee's view, different approaches must be explored to effectively expand the time available to provide an opportunity for an officer to be proficient and competent in both a service-specific and joint operational environment. To accomplish this will require innovation. For example, the Subcommittee notes that Congress recently provided authority to all the services for career flexibility. Specifically, the Navy is in the initial stages of executing a pilot program to allow officers a break in service to pursue other life events and then return and pick up where the officer left off. The need for such career flexibility reflected a potential future challenge of retaining highly qualified officers seeking to balance career and family needs. The Subcommittee observes that this approach could be equally applied to the officer development system to give officers an opportunity to pursue PME, JPME, or civilian advanced degrees, without concern over missing other critical career gates. Such an approach is but one idea that could be evaluated in a holistic examination of officer development and career progression.

Observation: There are multiple tensions among the officer's assignments necessary for service career development, the development of joint officers, the needs of the joint force, and professional military education.

***Finding*: The competing demands over the course of a 20- to 30-year career make it difficult to accommodate competing needs for the requisite education, training, and experience. The services have been seeking more flexible approaches, as exemplified by the joint qualification point system, for awarding joint duty credit and earning joint qualification.**

***Recommendation*: The services should review their officer development timelines from a holistic perspective to explore innovative avenues to develop their respective officer corps through education, training, and assignments or experience.**

Not only is there a tension between service and joint requirements in an officer's career, there are competing demands on the JPME system itself for producing different levels and types of expertise in joint matters. U.S. Joint Forces Command's 2010 *Joint Operating Environment* explores trends in warfare and their implications,[220] and, in doing so, provides the context for the *Capstone Concept of Joint Operations* (CCJO).[221] In examining these trends, the study forecasts the persistence of irregular warfare in addition to the possibility of major conventional conflicts, "over the next quarter century, U.S. military forces will be continually engaged in some dynamic combination of combat, security, engagement, and relief and reconstruction."[222] This complex future calls for military leaders who possess masteries of specific forms of joint expertise as well as those with a broad PME background. The CCJO establishes a number of requirements for officers with consequences for JPME.[223] These include:

- Improve knowledge of and capabilities for waging irregular warfare.
- Improve knowledge of and capabilities for security, engagement, and relief and reconstruction activities.
- Develop innovative and adaptive leaders down to the lowest levels.
- Develop joint commanders who are masters of operational art.
- Develop senior leaders who are experts not only in the operational employment of the joint force, but also in the development and execution of national strategy.[224]

The most recent version (2005) of the *CJCS Vision for Joint Officer Development*, which predates the CCJO, expresses its objectives for joint officer development. The CJCS vision consciously moved away from the specialist model of the joint staff officer envisioned by Goldwater-Nichols and the Skelton Panel and advocated educating "the largest possible body of joint officers suited for joint command and staff responsibilities."[225] Rather than viewing JPME I and II as preparation for an officer's first joint duty assignment, it describes a model where officers gain these joint qualifications by the colonel or Navy captain point in a career through various tracks, as opposed to an established sequential path.

In describing the new joint qualification system, it goes as far to say, "the JOD [Joint Officer Development] approach is fundamentally not building *specialists*, but inculcating jointness in *all* colonels and captains – a generalist approach."[226] Notwithstanding, the CCJO appears to call for competencies that may not necessarily be met by a model whose focus is producing generalists by the end of what would be a successful career for most officers. It is not clear, for example, how the generalist approach will satisfy the need for "senior leaders who are *experts* in the employment of the joint force" or "joint commanders who are *masters* of the operational art."[227]

Observation: There are competing demands on the JPME system for producing: joint "operators," joint staff officers, strategists, and senior leaders, among others.

***Finding*: Recognizing that the *Capstone Concept for Joint Operations* is the most recent guidance, the current *CJCS Vision for Joint Officer Development* does not adequately reconcile its generalist model with the requirement for specific joint competencies.**

***Recommendation*: In the subsequent revision of the *CJCS Vision for Joint Officer Development*, the Chairman of the Joint Chiefs of Staff should identify how the joint qualification system will fulfill the requirements established in the CCJO for various specific and specialized joint officer competencies.**

Shifting Competencies

In Subcommittee research on transition teams and provincial reconstruction teams in Iraq and Afghanistan, relatively junior officers were observed serving in positions requiring not only an understanding of some joint matters but interagency and multinational operations as well.[228] The question thus arose: "How soon is too soon to expose and introduce junior officers to joint concepts." Research showed that the Joint Staff levied substantial requirements at the pre- commissioning (e.g., *service academy*) and primary (e.g., *branch, warfare, and staff specialty schools*) levels. At the pre-commissioning level, the Learning Areas emphasize "knowledge of the basic U.S. defense structure, roles and missions of the other military services, the combatant command structure and the nature of American military power and joint warfare."[229] At the primary level, the CJCS Instruction on Officer Professional Military Education Policy (OPMEP) Learning Areas cover joint warfare fundamentals and joint campaigning to "prepare officers for service in joint task forces where a thorough introduction to joint warfare is required.[230] However, as service programs, the Joint Staff does not accredit pre-commissioning or primary schools.

Although not formally accredited by the Joint Staff, there was a requirement in the OPMEP for the Service Chiefs to report to the CJCS on their programs every three years.[231] However, this requirement was removed in the most recent edition of the OPMEP. The Joint Staff gave the rationale that this self-reporting method was not producing useful evaluations of the joint instruction. The Joint Staff reported to the Subcommittee that it is in the process of developing a more suitable assessment tool.[232]

During visits to the institutions, the Subcommittee found that the service academies, Squadron Officers College, and Expeditionary Warfare School were all aware of the OPMEP requirements and could identify where they were being addressed in their curricula. Apart from the service academies and the two primary-level schools, the process of assessing the other pre-commissioning programs becomes more complicated. Given the number of institutions involved, which includes Reserve Officer Training Corps (ROTC) programs and officer direct-entry (officer training school/officer candidate school) locations for all three departments, it may not be practical to conduct the Process for Accreditation of Joint Education (PAJE) or even PAJE-like evaluations of the joint instruction at each one. What appears feasible is a PAJE review for joint content of the services' guidance for Reserve Officer Training Corps and officer direct-entry programs.

***Finding*: With the elimination of self-assessment reporting, there is no evaluation process to monitor compliance of pre-commissioning- and primary-level PME schools with the JPME requirement in the OPMEP.**

***Recommendation*: The Department and CJCS should expedite development of an evaluation process to ensure that the services are effectively teaching pre-commissioning and primary joint education. Additionally, the Department and CJCS should consider that this process include a review of the joint curricular guidance that the services give to their pre-commissioning and primary-level schools to include ROTC and officer direct entry programs.**

Beyond the most junior levels, many intermediate-level school students said that they needed earlier exposure to operational planning. This was especially true for Navy, and in some cases Air Force students, who felt that their Army and Marine Corps counterparts were better prepared.[233] This may have been due, to some extent, to the fact that some intermediate-level students were uncomfortable with the operational planning process that they were seeing for the first time. One rationale for mandating an earlier introduction to operational planning is the previously-mentioned number of O-4 (major and Navy lieutenant commander) billets on combatant command and joint task force staffs being filled by O-3s (captains and Navy lieutenants).

***Finding*: There is an increasing need for additional joint subject matter to be taught at the primary PME level, especially joint planning and execution processes.**

***Recommendation*: The Department and CJCS should review both the OPMEP requirements at the primary level and joint content requirements for junior officers in the current operating environment. They should consider adding familiarity with joint planning and execution processes.**

In addition to the need for ensuring that joint subject matter is being taught at the pre-commissioning and primary levels, which represents a shift of some joint competencies to earlier in an officer's career, there has also been some homogenization of subject matter among the senior colleges brought about largely as a result of common requirements in the OPMEP. This is particularly evident in the teaching of the levels of strategy.

The Skelton Panel envisioned that "the National War College [National] should decrease the amount of time devoted to national military strategy and become a center for the study of national *security* strategy."[234] The Panel also recommended that the service war colleges focus on national military strategy and theater-level campaigns or operations. The rationale for this was based on the Panel's observation that there was a lack of depth to the curricula at the senior PME schools, a deficiency that was characterized by Professor Williamson Murray as "the Pecos River approach – a mile wide and an inch deep."[235]

The concentration on different levels by National and the service war colleges was to allow a sharpening of focus, allowing their respective curricula to be treated in greater depth.[236] In the intervening years however, this division of labor among the schools appears to have been overtaken by both a more complex security environment and by law. The original rationale for it may no longer exist.

The distinction between military power and other instruments of national power is not as sharp as it was in 1989 as the Cold War was drawing to a close. In irregular warfare, for example, the military acts in many capacities usually associated with other departments, agencies, industry, and non-governmental organizations. Additionally, since the Congress granted the senior service schools JPME II status, they are required by law[237] and the OPMEP[238] to teach national security strategy.

In the Subcommittee's review of the PAJE team reports and in visits to the colleges, no demonstrable problem with the senior schools teaching both national security strategy and national military strategy was evident. The senior schools assign the appropriate weight among the curricula areas set out in law which include, among others, national security strategy, national military strategy, and theater strategy and campaigning.[239]

***Finding*: A major purpose of this study was to understand the extent to which the recommendations of the Skelton Panel have been implemented. The Subcommittee observed that the Panel's recommendation, that National focus its curriculum on national security strategy, is still explicitly provided for in the OPMEP and continues to be practiced. It also found, however, that the rationale for the senior service schools focusing only on national military strategy is no longer as convincing as it once was.**

Civilian Graduate Schools

All the services have programs to send officers to civilian graduate schools, apart from those who are preparing for military faculty positions at various schools. Typically, these are select officers chosen to study one to two years at top-tier schools in place of attending the PME institutions associated with their rank. Although reconcilable, the Subcommittee observed tension between the point of view that civilian graduate education is better suited for imparting critical thinking skills and the view which holds that there are unique benefits to studying in the multiservice, and increasingly interagency, environments at the PME institutions.

The Skelton Panel considered international relations, political science, economics, and history as the "core components" of national security strategy, and by extension of PME. Potential strategists developing competence in these disciplines, in the Panel's view, could not wait until an officer would be eligible to attend a war college. The Panel recognized that earlier study in "prestigious" civilian graduate schools was the most practical means to continue acquiring knowledge in these fields. At that time, fewer of the service and joint intermediate, or senior schools granted master's degrees, but the Panel accepted that the Naval Postgraduate School and the Air Force Institute of Technology also might be able to contribute to this graduate-level education in that at least the former offered degrees in national security affairs.[240]

The Skelton Panel cited Admiral William Crowe, CJCS at the time, and retired General Andrew Goodpaster, both of whom earned doctorates from Princeton while in the military, as proponents of civilian graduate programs in these national security affairs-related fields.[241] Some of today's senior officers expressed similar views regarding this experience. General George Casey, the current Chief of Staff of the Army, who has a master's degree in international relations from the University of Denver, spoke highly of the formative nature of his studies there.[242] He mentioned his initiative, aimed in part at retention, which offered the incentive of funded civilian graduate education for up to 400 West Point and ROTC graduates as an inducement to serve past their initial obligation.[243] General David Petraeus, commander of CENTCOM and a Princeton University Ph.D., referred to the advantages of future leaders and commanders getting out of their "intellectual comfort zones," stating that "few if any experiences we can provide within our military communities are as intellectually stimulating, challenging, or mind-opening as a year or two at a civilian graduate school."[244] He recalled from his own experience that "[d]ebates we imagined to be two-sided turn out to be three-, four- or more sided."[245] Retired Lieutenant General David Barno commanded U.S. and coalition forces in Afghanistan from 2003 to 2004. Looking back at his PME experience he recalled:

> There is no substitute for a civilian graduate degree to sharpen the thinking of our officers as they move up through the ranks and they become senior officers. That helped me more—my graduate schooling at Georgetown University as a captain helped me more—than perhaps any other developmental experience at the strategic level.[246]

Professor Williamson Murray noted to the Subcommittee that, in terms of intellectual development, captain (Navy lieutenant) was the right point in an officer's career for this experience. He stated that, "the crucial point, I think, is the captain level. If you look at people like [Lieutenant General] Don Holder and Petraeus and various other individuals who have gotten the mark as first rate strategists, they have gotten that mark really in terms of beginning to fill their gas tank at the captain level."[247]

***Finding*: The intellectual development of officers, especially in critical thinking skills, is facilitated by assignment to civilian graduate education programs at top-tier universities relatively early in their careers.**

While the Skelton Panel recognized that officers must look outside the PME system to develop certain competencies and saw the role of education outside the system as necessary, it emphasized that it was not a substitute or replacement for PME.[248] Some experts who testified to the Subcommittee echoed this view that civilian education and PME for officers is not an "either- or" proposition.[249] In response to a question on the existential value of PME, and in reference to a suggestion that the war colleges be closed down, Dr. Janet Breslin-Smith, retired professor and former department head at National, spoke of the distinctive character of the in-residence experience at her former institution and other PME schools:

> But at least in the case of the War College, and I think a number of the other schools here, the type of interaction that happens in the classroom—again, going back to Eisenhower's image of this—is exactly what we talk about these days. How do we get a total national security team, [U.S. Agency for International Development], State, and military officers to be able to work together, understand each other's culture, before they are in the field together? So the type of education that goes on wouldn't be accomplished if you have everybody going to a university taking poli sci or international relations classes.[250]

***Finding*: There are unique benefits of PME relative to civilian education programs. Civilian education programs at the top-tier schools, however, complement PME.**

Commentators emphasized the mutual benefit in sending military officers to civilian graduate programs. Professor John Williams of Loyola University in Chicago testified specifically in terms of civil-military relations:

> I want our elite military officers meeting the brightest, most elite civilians, and I want them interacting with each other. I want them to put a human face on one another. I want the military to get how civilians think, and I want the civilians to get how the military thinks and not be lured into stereotypes. I think it would be beneficial for civil-military relations, especially since they don't really have to come together on many occasions.[251]

Observation: More military officers' interaction in civilian academic environments benefits both groups of people and institutions.

In addition to the enhanced broadening experience of graduate school, there is a degree of mastery of a discipline and a development of higher-order thinking skills that come from the requisite examinations, focused research, and writing requirements associated with earning a doctorate in a field that go beyond that attained in gaining a master's degree. This raises the question of the costs and benefits to the services of developing small cadres of officers with doctorates in fields related to national security. General Petraeus and Admiral James Stavridis are two who have been mentioned in this regard as preeminent military strategic thinkers who were aided in their intellectual development by doctoral study at top universities.

All of the services, with the exception of the Marine Corps, have established programs for sending officers for doctoral study. This service-sponsored advanced education largely supports faculty positions at the service academies and military graduate schools with a small number specializing in security studies-related disciplines. Accordingly, this graduate work is typically in preparation for subsequent teaching in these fields as opposed to assignment to strategy-related billets on high-level staffs or to strategic leader positions. The Marine Corps, which largely relies on the Navy to support teaching positions for officers and civilian professors at the Naval Academy and Naval Postgraduate School, reported that they do not have the same requirement in this regard as the other services.[252]

There is still the question of whether the services would benefit from small number of officers more deeply steeped in academic disciplines related to security studies, apart from those in academic positions. Although witnesses in Subcommittee PME hearings stated that the additional numbers within each service would be few,[253] Admiral Michael Mullen, the current CJCS, suggested the number might be more substantial than currently exists.[254] Professor Murray suggested to General James Mattis, who at the time was the commanding general of the Marine Corps Combat Development Center, which oversees the Marine Corps University, that his service should experiment with a trial program for six or seven captains a year to pursue a doctorate in strategic studies at "elite" civilian universities.[255]

***Finding*: The services primarily send officers to graduate school in security-related fields to prepare them for teaching assignments.**

***Recommendation*: The services should sponsor a number of junior/company grade officers for Ph.D.s in strategic studies including history, political science, international relations, and economics at top-tier civilian institutions, beyond that needed for faculty positions at academic institutions, in order to build a cadre of strategic thinkers for the operating forces and higher-level staffs.**

Apart from civilian doctoral programs, some PME institutions are proposing to fill some of this need on their own. Prior to the passage of the FY 2009 NDAA, only the Naval Postgraduate School was specifically authorized to award doctoral degrees. Language in the bill that year standardized the degree-granting power of the in-residence PME institutions and invested them with the very broadly-written authority to "confer appropriate degrees upon graduates who meet the degree requirements."[256]

As a result of this change, the Air University is moving forward with plans for developing a Ph.D. for airpower strategists.[257] The Air Force does not envision this as a faculty development source, but rather preparation for senior staffs and command positions.[258] Some Industrial College of the Armed Forces (ICAF) faculty also have a concept for instituting a doctoral program. This proposal envisions a follow-on year for selected candidates, military and civilian, from all the senior- level schools at a newly established NDU School for Advanced Strategic Planning and Strategy.[259]

While both of these proposals would have civilian elements, neither replicates the setting described by General Barno and others, with military officers being placed outside of their comfort zones, having civilian peers challenge their perspectives, who may have no particular familiarity or experience with the military, at a top-tier graduate school. The students participating in the ICAF and School of Advanced Air and Space Studies (SAASS) programs would have had shared formative military experiences and would be predominately senior lieutenant colonels (Navy commanders) and colonels (Navy captains). As a result, these studies would not occur at the captain (Navy lieutenant) point, or at approximately 25-30 years of age, the point in intellectual development when some think an officer is most open to developing critical thinking skills.

Observation: There are at least two proposals to create military doctoral programs at PME institutions including ICAF and SAASS. This is currently allowed in legislation.

***Finding*: Although these doctoral program proposals are not sufficiently advanced to evaluate with regard to the extent that they fulfill service-specific needs, neither has a sufficient civilian academic component to create the type of strategic thinkers that the military needs or the Skelton Panel envisioned.**

***Recommendation*: That the services do not view internally-administered doctoral programs as substitutes or replacements for existing civilian graduate-education programs.**

Developing Strategists

> *"[B]y its nature strategy is more demanding of the intellect and perhaps imagination than any structurally more simple activity – policy, operations, tactics, or logistics for prominent examples. Excellence in strategy requires the strategist to transcend simple categories of thought."*[260]
>
> **Dr. Colin S. Gray**

In reviewing the PME system, the Subcommittee made an effort to assess: how the military identifies and cultivates strategists and the degree to which PME develops strategic thinkers and decision makers. The Skelton Panel established a valuable precedent in its landmark report on PME. The Subcommittee considered the Skelton Panel's approach to teaching strategy and many of its recommendations for developing strategists as appropriate bases for evaluation.

The Skelton Panel recognized that talent, experience, and education are the key elements for cultivating strategists.[261] As to the educational aspect, it declared: "Original and independent strategic thinkers can be shaped and molded by a variety of educational experiences, but PME must be an important part of these diverse experiences."[262] The Skelton Panel also emphasized that: "A defense establishment that seeks to encourage the development of strategists must ensure that this scarce national resource is used in the most effective manner possible."[263] The Skelton Panel urged the military to identify promising strategists as early in their careers as possible, and, in order to develop them efficiently thereafter, it recommended synergizing the systems for officer selection, assignment, and education.[264] It concluded that "each service should have a personnel management system to develop, monitor, and assign officers to service and joint billets that would benefit from an officer with expertise in strategy."[265] The Skelton Panel also recommended that the CJCS should "ensure that the need of joint, departmental, and national-level organizations for strategists is met," and that billets "requiring strategists should be so designated on the joint duty assignment list – including some critical joint duty assignment positions."[266] Finally, the Skelton Report asserted that a "conscious effort" should be made to develop and designate certain joint specialty officers (now joint qualified officers) as strategists.[267]

The Subcommittee conducted its assessment with an appreciation for strategists as a scarce, yet vital, human resource and for PME as an irreplaceable part of their educational development.

The Demand for Developing Strategists

Prior to assessing how the military identifies and cultivates strategists, the Subcommittee considered why, and for what purposes, the Department develops them. Strategy has been defined in numerous ways, and the exercise of defining strategy has often been signified by vigorous debate and lack of consensus. Modern views tend to regard strategy as a purposeful combination of ends (or objectives), ways (or courses of action), and means (or instruments for achievement).[268] Strategizing involves balancing priorities, evaluating options, and making calculated choices with respect to each of these interrelated variables to produce an acceptable effect.[269] The Department currently defines strategy as: "A prudent idea or set of ideas for employing the instruments of national power in a synchronized and integrated fashion to achieve theater, national, and/or multinational objectives."[270] Strategists are expected to carefully formulate broad, goal-oriented theories for military action and, whenever necessary, put them into practice.[271] The Skelton Panel took a similarly pragmatic view in 1989 when it defined strategy as "the link that translates power into the achievement of objectives."[272]

Because American national objectives, and the laws, policies, and executive decisions that govern their pursuit, are ordained by the political process, military strategists must translate options for military action, in coordination with those associated with various other forms of statecraft, into demonstrable results.[273] Professor Colin Gray describes the relationship as follows:

> Strategy, after all, is the bridge connecting the threat and use of force with policy or politics. The strategist needs to understand what is tactically and operationally possible in all geographical environments, what success or failure in each environment (or functional dimension) contributes to performance in the other environments, what that means for military performance writ large, and what general military performance means for policy (and vice versa).[274]

The functional dimensions of strategy are often broken down according to scope. The Skelton Panel categorized strategy into three nested and mutually influential subsets: national security strategy, national military strategy, and operational art.[275] National security strategy was defined as: "The art and science of developing and using the political, economic, and psychological powers of a nation, together with its armed forces, during peace and war, to secure national objectives."[276] Military strategists contribute to the national security strategy through the exercise of national military strategy. It was defined as: "The art and science of employing the armed forces of a nation to secure the objectives of national policy by the application of force or the threat of force." Operational art was defined as: "The employment of military forces to attain strategic goals in a theater of war or theater of operations through the design, organization and conduct of campaigns and major operations."[277] Although operational art is frequently considered to be a separate and intermediate level of warfare, between the strategic and tactical levels, it has also traditionally been considered the level at which strategic objectives can be realized by large, joint formations.[278]

Joint and interagency activities are increasingly regarded from a strategic perspective. The current *Capstone Concept for Joint Operations* (CCJO) stresses that joint force commanders must think strategically. The CCJO explains:

> While operational expertise is essential, it is not enough. In a future requiring integrated national effort, joint force commanders cannot afford to focus narrowly on achieving assigned operational objectives, but must contribute to the development of strategic objectives as well. They must be knowledgeable about the use not only of the military instrument, but also all the other elements of national power, how those elements interact with military force, and how they ultimately might supplant the need for military force. Development of that broader strategic understanding must begin early in the military education process and continue throughout every military officer's professional development.[279]

As for enabling strategic success within a complex security environment, the *Joint Operating Environment* enunciates the demand for strategic agility amid uncertainty.[280] The *Joint Operating Environment* counsels American strategists that they will confront "the conundrum of preparing for wars that remain uncertain as to their form, location, level of commitment, the contribution of potential allies, and the nature of the enemy" and that these "strategists will have to prepare to work in an environment where the global economic picture can change suddenly, and where even minor events can cause a cascading series of unforeseen consequences."[281] Strategists must develop discerning and adaptive capacities for understanding conflict, in its various forms, as it continues to evolve amid the innumerable complexities and volatilities of the security environment. Strategic understanding must also account for the motivations and objectives of other actors, whether they are allied with or opposed to those of the United States, and it must facilitate the translation of practicable strategic objectives into effective operational and tactical performance.[282] Perhaps most importantly, strategic understanding should be developed in a manner that heeds lessons learned and that can anticipate significant change, so that costly errors can be minimized in the unfortunate event of actual conflict. Strategists should avoid being too predictive.[283] The *Joint Operating Environment* guardedly notes: "We will likely not call the future exactly right, but we must think through the nature of continuity and change in strategic trends to discern their military implications to avoid being completely wrong."[284] Nevertheless, the demand for military strategists requires talented individuals who have been cultivated through their education, training, and experience to recognize and address difficult strategic issues as they emerge and develop. Strategists must be able to approach strategic issues with prudential foresight, rather than with disabled or misapplied notions as to the nature of a given conflict.

Therefore, a concerted developmental effort for producing strategists is still required. The *Joint Operating Environment* asserts, "If we expect to develop and sustain a military that operates at a higher level of strategic and operational understanding, the time has come to address the recruiting, education, training, incentive, and promotion systems so that they are consistent with the intellectual requirements for the future Joint Force."[285] Williamson Murray argues that the military should promote intellectual agility by rewarding outstanding academic performance with command and expanded educational opportunities in the areas of strategy, military history, and regional studies. He testified: "Such changes would demand a fundamental shift in the cultural patterns of the services, particularly in their personnel systems as well as their career patterns."[286]

Observation: The security environment demands that cultivation of a broad strategic understanding and the promotion of intellectual agility amid strategic change begin early in an officer's military education and continue throughout his or her professional development.

The Attributes of a Strategist

The Skelton Panel distilled the attributes of a strategist to four. It asserted that strategists must be analytical, pragmatic, innovative, and broadly educated in domestic and international political, technological, economic, scientific, and social trends.[287] The Skelton Panel acknowledged that few officers possess all four attributes, noting: "It is rare to find individuals capable of a high degree of conceptualization and innovation – the attributes that most distinguish the theoretical from the applied strategist."[288] The Skelton Panel concluded that few theoretical strategists are needed to be effective, and that the PME system should endeavor to develop a small number of them. Yet, the Skelton Panel also perceived that numerous applied strategists equipped with practical problem-solving skills (as distinguished from those for conceptualization and innovation) are needed. The Panel further reasoned that this larger grouping is easier to cultivate. The Skelton Report proposed that "the goals of the PME system with respect to strategists should be two-fold: (1) to improve the quality of strategic thinking among senior military officers and (2) to encourage the development of a more limited number of bona fide theoretical strategists."[289]

Strategists at the Naval War College.

Today, the requirement for both theoretical and applied strategists persists. Ideally, individuals possessing all of the attributes lauded by the Skelton Panel would succeed to positions of senior leadership and command authority.[290] This goal places a heavy premium on combining the rare innovative attribute (as enhanced by extensive education) with that for pragmatic and resourceful military performance.[291] Professor Gray argues that national security "requires of its creative strategist(s) the ability to turn brilliant insights into effective command performance."[292] He continues, "In other words, it is not sufficient to educate strategists who know what should be done, or at least what might with great boldness be attempted. Also, there is an absolute requirement for a few, fortunately probably only a very few, strategists who are people of action as well as creative thought."[293] Rare strategic ability is best invested in the unique combination of authority, responsibility, and accountability that constitutes command at the strategic level of warfare. Undoubtedly, a well-educated cohort of strategically-minded analysts and implementers will continue to be needed to support myriad strategic applications. However, the military should identify and cultivate the most creative and innovative of its strategic thinkers for the significant intellectual and leadership challenges of command decision-making at the strategic level.

Observation: The military requires a small number of creative and innovative theoretical strategic decision-makers and a larger number of strategic analysts and implementers.

Joint Perspectives on Developing Strategists

The CJCS and the Joint Staff influence the development of strategists through JPME, management of the joint PME schools, and joint experience. The services contribute to the identifying and cultivating of strategists by other means, including through service-specific PME. The Joint Staff does not identify or monitor strategists or potential strategists, unless an officer has either: 1) attended a joint PME institution (i.e., the Joint Advanced Warfighting School); or 2) completed a tour in a "joint strategist-annotated" position. The Joint Staff can track these officers through the Joint Duty Assignment Management Information System (JDAMIS) and the Electronic Joint Manpower and Personnel System (eJMAPS). No formal strategic sub-specialty is identified in the Joint Qualification System, which currently governs the qualification levels and advancement of joint qualified officers. The Joint Staff does not monitor the progress of strategists educated at the services' PME institutions (unless they subsequently complete joint strategist-annotated assignments) or strategists who only serve in service-specific strategy billets. These personnel tracking functions are reserved for the services.[294]

The OPMEP recognizes that identifying and developing "officers with the capacity for strategic thought" is an educational responsibility of military leaders at all levels.[295] It notes: "PME provides the education needed to complement training, experience, and self-improvement to produce the most professionally competent (strategic-minded, critical-thinking) individual possible."[296] The OPMEP specifically states that the PME system should produce:

> 1) Strategically minded officers educated in the profession of arms who possess an intuitive approach to joint warfighting built upon individual Service competencies. Its aim is to produce graduates prepared to lead the Capstone Concept of Joint Operations [CCJO] envisioned force within a multi-Service, multi-agency, multi- national environment and able to participate in and contribute to informed decision- making on the application of all instruments of national power.
>
> 2) Critical thinkers who view military affairs in the broadest context and are capable of identifying and evaluating likely changes and associated responses affecting the employment of U.S. military forces. Graduates should possess acuity of mind at the highest level; gained as a result of a continuum of learning across a lifetime.
>
> 3) Senior officers who, as skilled joint warfighters, can develop and execute national military strategies that effectively employ the Armed Forces in concert with other instruments of national power to achieve the goals of national security strategy and policy in the air, land, maritime, and space physical domains and the information environment (which includes cyberspace).[297]

These OPMEP assertions appear responsive to relevant demand signals for strategists. They stress the fundamental role education plays in developing strategic thinkers and decision makers, and they accept strategy as a creative and pragmatic activity that must adapt to opposition, change, and uncertainty.[298]

The OPMEP supports early emphasis on the importance of strategic perspective. It states: "PME needs to continue to build an officer that understands the strategic implications of tactical actions and the consequences that strategic actions have on the tactical environment."[299] The Joint Staff also supports nurturing strategists throughout their careers, as they become more educated, trained, and experienced in strategic matters. It especially advocates identifying and cultivating strategists through senior-subordinate mentorships.[300] The Joint Staff regards these mentorships as vehicles for continued learning, advantageous assignments, and career viability.[301]

Finding: **The Joint Staff identifies or monitors strategists only if they have either: (1) attended a joint PME institution; or (2) completed a tour in a joint strategist-annotated position. The Joint Staff can track these officers through the Joint Duty Assignment Management Information System (JDAMIS) and the Electronic Joint Manpower and Personnel System (eJMAPS). No strategic specialty is formally identified in the Joint Qualification System, which currently governs the qualifications and advancement of JQOs.**

Recommendation: **The Joint Staff should evaluate the demand for joint strategist-annotated billets during its JDAL validation process. The Joint Staff should also develop and maintain a centralized registry of all designated military strategists regardless of the manner in which they may have earned JPME credit or whether they have previously served in a joint strategist-annotated billet.**

Service Perspectives on Developing Strategists

The Skelton Panel reported that only the Army and the Navy possessed personnel systems to identify officers with educational credentials in strategy and that only the Navy utilized a system for monitoring and assigning officers to strategy-related billets. At that time, neither the Air Force nor the Marine Corps tracked or monitored strategists on the basis of strategy-related education or experience.[302]

The Army

The Skelton Report observed that the Army recognized strategic expertise by assigning a skill identifier to qualified officers in addition to their primary (i.e., combat arms) and secondary (e.g., personnel or operations) specialties. This skill identifier was awarded on the basis of: 1) having achieved a master's degree in a social science; and, 2) having received requisite intermediate PME at the Army Command and Staff College (Army CGSC) with a concentration in strategy; or having received intermediate PME at another joint or service school and having served for at least 12 months in a strategy-designated billet. However, the Skelton Report also noted that the Army had no formal program for monitoring the careers of those officers with a strategy skill identifier for purposes of assignment.[303]

The Army continues to assign a strategy-related skill identifier to qualified field-grade officers. The 6Z (Army Strategist) additional skill identifier may be earned by completing either: (1) the Defense Strategy Course, which is a six-month U.S. Army War College (USAWC) distance learning program focused on national security strategy; or (2) the 6Z elective offered at the Army CGSC's 10-month Advanced Operations Warfighting Course. However, the 6Z designation is now considered to be a primer for the Army's broader effort to cultivate strategists, the Strategic Plans and Policy Functional Area 59 (FA 59) program.[304] According to the Army pamphlet on Commissioned Officer Professional Development and Career Management, FA 59 officers are:

> ...warfighters who provide the Army with a highly trained cadre specializing in the development and implementation of national strategic plans and policies; theater strategy and campaign planning; and the evolution of concepts and doctrine for employing military forces at the operational and strategic levels of warfare.[305]

The Army relies on FA 59 officers to: 1) conduct strategic appraisals that are responsive to "adaptive adversaries, shifting ends, and complex situations;" 2) develop strategic plans or recommendations that can translate operational means into agreeable ends; 3) integrate inter-service and interagency capabilities and contributions; and, 4) teach curricula to support education in "military theory, the strategic arts, and national security strategies and policies."[306] These descriptions support the observation that FA 59 officers are limited to analytical and advisory functions in support of separate command authorities. FA 59 officers are not eligible for command at the battalion level or above.[307] The Army maintains that career timing restraints combined with extensive educational requirements effectively preclude FA 59 officers from higher command opportunities. Consequently, promotion beyond the rank of colonel is extremely unusual. Only one officer has been selected for promotion to brigadier general in the history of the FA 59 program.[308]

Approximately 350 Army officers are currently designated as FA 59 officers. Roughly 65 percent of these officers serve in billets reserved for FA 59 officers. About 20 percent of them serve in billets that are open to other specialties, and the remaining 15 percent are serving in education billets as either students or trainees. Approximately 10 to 15 officers enter the FA 59 program each year.[309] Officers typically enter the program as Army senior captains or junior majors with "history, policy, business, and economics academic backgrounds."[310] Officers with graduate degrees in these disciplines and those with experience in plans and policy assignments are preferred applicants.[311]

Specialized PME contributes significantly to FA 59 officer development. All FA 59 officers may enroll in the Defense Strategy Course, all are expected to earn the 6Z additional skill identifier before rising to senior major, and all must attend a 14-week ILE course in Army-specific assignments and operations in joint interagency, intergovernmental, and multinational (JIIM) environments. Select FA 59 officers attend the College of Naval Command and Staff (CNCS) to earn masters' degrees in strategic studies. PME for some FA 59 officers includes the Army's Basic Strategic Arts Program at Carlisle Barracks, which helps junior majors "bridge the gap between their tactical/operational background and the challenges of operating at the grand-strategic and theater- strategic levels of war and policy."[312] Some FA 59 officers may attend the Advanced Military Studies Program (AMSP) at Fort Leavenworth or an equivalent program at another service or joint PME school. The AMSP curriculum supports "integrated study of military history, military theory, and execution-based practical exercises" to confer a master's degree in Military Arts and Sciences. These students may also choose to pursue a master's degree in security studies from a civilian institution partnered with the Army CGSC.[313]

FA 59 officers are urged to attend the Joint Advanced Warfighting School as senior-level education, because its curriculum underscores strategic planning, strategic appraisal, and joint and interagency integration. Colonels and lieutenant colonels attending the USAWC are also encouraged to apply to its Advanced Strategic Arts Program, and exchange officers attending joint and other senior service schools are encouraged to pursue those institutions' similar strategic study programs. FA 59 officers are also invited to pursue certain civilian graduate programs. Junior FA 59 officers may attend the Army Harvard Strategist Program or pursue a master's degree in public policy, history, political science, or international relations at another civilian school. Senior FA 59 officers may attend Seminar XXI at the Massachusetts Institute of Technology with governmental and non- governmental civilians. The seminar examines policy issues through case-study exercises that feature competing national and international perspectives. FA 59 officers are also encouraged to earn civilian doctorates in policy-related fields.[314]

FA 59 assignments are predominately Army-centric, but FA 59 officers also serve in joint duty assignments. Approximately 18 percent of all FA 59 officers currently serve in COCOM billets.[315] An FA 59 career manager at the Army Human Resources Command evaluates each FA 59 assignment candidate for unique combinations of operational experience and education.[316] FA 59 officers serve in: plans and policy staff billets within Army divisions, corps, or theater armies; Army commands; COCOMs; multi-national headquarters; joint task forces; and, staff and faculty positions at PME institutions.[317] Only FA 59 colonels who have experience at a major Army or joint command may be recommended for interagency positions.[318]

Finding: **The Army developed the FA 59 program, which supports strategic field-grade specialists in performing advisory services in Army, joint, and interagency billets at theater and national strategic levels. This program affords ample opportunities for educational development, but it precludes command opportunities. Because participation in the FA 59 program severely restricts viability for promotion beyond the rank of colonel, the Army has limited ability to add educated and experienced strategists to its general officer ranks.**

The Navy

The Skelton Report observed that the Navy also recognizes strategic expertise by awarding subspecialty codes to line officers in addition to their primary warfare specialty (e.g., surface warfare, aviation, or subsurface warfare). These subspecialty codes were awarded on the basis of having: (1) earned a graduate degree in a strategy-related discipline; (2) acquired experience in a strategy-related billet (e.g., the Office of the Deputy Chief of Naval Operations for Plans, Policy and Operations); or (3) a combination of board-approved education and experience. Designated naval officers were eligible for assignment to strategic-level positions at major commands (e.g., the National Security Council, the Office of the Secretary of Defense, the Joint Staff, the Office of the Chief of Naval Operations (Plans, Policy, and Operations), and a multitude of joint and fleet commands), and all strategy-related assignments were managed by a subspecialty coordinator.[319]

The Navy continues to manage strategists and potential strategists in this manner. The Bureau of Naval Personnel manages the Navy Subspecialty System (NSS), which monitors officers with strategy-related education and experiences. The NSS tracks officers' education and experience credentials and matches them with specific requirements for strategy-related billets in the assignment process.[320] As of March 2010, the NSS tracked 5,170 subspecialty codes among naval officers (including 127 subspecialty codes awarded to 105 flag officers) in support of 687 billets reserved for the following strategy-related subspecialty codes:[321]

These subspecialty codes are conferred only on the basis of demonstrable education and experience. For example, subspecialty code 2000 (National Security Studies, General) requires officers adept in:

- Formulating and/or evaluating national/international policy and/or strategy. This includes but is not limited to naval doctrine, joint strategy and operational planning.
- Theoretical and practical understanding of national military capabilities, command structure, joint doctrine, intelligence sources, multi-national sources at the operational and tactical levels of war, joint planning and execution.
- Use of analytical tools, threat analysis and research methods to evaluate the effect of local and/or regional political, cultural, and security aspects of DOD programs and objectives. This includes a working knowledge of state of the art analytical tools (e.g., assessment, forecasting, gaming, and/or simulation) and/or the intelligence cycle and research methodologies.[323]

Subspecialty Code	Subspecialty Title
2000	National Security Studies, General
2101	Regional Studies: Middle East, Africa and South Asia
2102	Regional Studies: Far East and Pacific
2103	Regional Studies: Western Hemisphere
2104	Regional Studies: Europe, Russia and Associated States
2200	Regional Intelligence, General
2201	Regional Intelligence: Middle East, Africa and South Asia
2202	Regional Intelligence: Far East and Pacific
2203	Regional Intelligence: Western Hemisphere
2204	Regional Intelligence: Europe, Russia and Associated States
2400	Strategic Intelligence
2500	Special Operations/Low Intensity Conflict
2600	Homeland Security and Defense
2700	Security Studies: Stability and Reconstruction

[322]

These subspecialty codes are conferred only on the basis of demonstrable education and experience. For example, subspecialty code 2000 (National Security Studies, General) requires officers adept in:

- Formulating and/or evaluating national/international policy and/or strategy. This includes but is not limited to naval doctrine, joint strategy and operational planning.
- Theoretical and practical understanding of national military capabilities, command structure, joint doctrine, intelligence sources, multi-national sources at the operational and tactical levels of war, joint planning and execution.
- Use of analytical tools, threat analysis and research methods to evaluate the effect of local and/or regional political, cultural, and security aspects of DOD programs and objectives. This includes a working knowledge of state of the art analytical tools (e.g., assessment, forecasting, gaming, and/or simulation) and/or the intelligence cycle and research methodologies.[323]

Subspecialty code 2000 may be "earned through successful completion of an experience tour in a political science/security affairs or joint/operational intelligence billet, or graduate level education."[324] The Navy leverages opportunities within and beyond PME. Graduate degree programs must be properly accredited in a relevant field (e.g., strategy, national security affairs, international relations, political science, intelligence).[325] The Naval War College offers a master's degree in national security and strategic studies.[326] Approved civilian institutions include: Georgetown University, Harvard University, the School of Advanced International Studies at Johns Hopkins University, Stanford University, and the Fletcher School of Law and Diplomacy at Tufts University.[327] Naval officers may also earn a qualification designator by researching a strategic issue through the Naval War College's selective Advanced Research Program or either of its Mahan or Halsey scholarship programs.[328]

Naval War Gaming Today.

The Navy's reliance on subspecialty designations supports the primacy of an officer's qualifications in an established naval warfare community. This is consistent with the Navy's culture of seagoing command. The Navy identifies preparation for "strategic leadership" as a core function of the Naval War College curriculum.[329] Therefore, a designated Navy strategist is eligible for command in his or her community, which allows for productive mergers of strategic expertise and command authority. The challenge in this approach lies in synchronizing the education and experiences necessary for developing a strategic perspective with those necessary for successful command at every level. Both objectives are demanding, and they are not easily joined within the time constraints of a single career. In the Navy, competing operational and technical proficiency requirements can take precedence over educational opportunities, including those in strategic studies, when timing is constrained.[330]

***Finding*: The Navy continues to assign strategy-related subspecialty designations to line officers in addition to their primary warfare designations on the basis of: (1) having earned a graduate degree in a strategy-related discipline; (2) having served in a strategy-related billet; or (3) having an approved combination thereof. Designated naval officers are assigned to strategy-related billets at major fleet, joint, and interagency commands. The Navy's reliance on subspecialty designators supports primacy of an officer's warfare qualifications. As a result, Navy strategists are eligible for command and are afforded opportunities to merge strategic expertise with the authority to make command decisions. The challenge in this approach lies in synchronizing the education and experiences necessary for developing a strategic perspective with those necessary for successful command at every level. In the Navy, competing demands can take precedence over educational opportunities in strategy.**

The Air Force

The Air Force does not maintain a distinct system for cultivating strategists. Instead, the Air Force assigns a Development Team for each of its occupational communities. Each Development Team assesses the education, training, and experience of officers within its assigned community and helps to guide members in gathering credentials appropriate to the service's needs. Development Teams may identify potential strategists and recommend them for further training and educational opportunities (e.g., in-residence PME, advanced functional training, and graduate education degrees), position types (e.g., flight commander, division chief, instructor, special duty), or assignments to strategic-level organizations (e.g., the Joint Staff, Air Staff, Air Force Major Commands). The Air Force monitors graduates of the School of Advanced Air and Space Studies as having received a professional education in strategy.[331]

***Finding*: The Air Force screens its officers for applicable credentials across each of its functional communities, but it does not specifically identify or cultivate strategists independently.**

The Marine Corps

The Marine Corps does not employ a formal identification or tracking mechanism for strategists within its personnel system. Rather, it broadly assesses officers for "requisite skills and/or potential to serve properly" in matching qualified Marines with suitable assignments.[332] The Marine Corps maintains: "Taken in context, the goal is that all Marines evolve into 'strategists' over the course of their careers and as appropriate to their training, education and assignments."[333]

***Finding*: The Marine Corps does not employ a formal identification or tracking mechanism for strategists within its personnel system.**

***Finding*: The Army and Navy have relatively advanced systems for cultivating strategists, while the Air Force and Marine Corps systems remain relatively underdeveloped.**

***Recommendation*: Each of the services should carefully review and further develop relevant processes for identifying and cultivating strategists. In doing so, they should optimize the development of qualified strategic decision-makers, in addition to strategic analysts or advisors, and they should endeavor to balance academic achievement in a strategy-related discipline with command experience.**

Strategic Studies Programs within the PME System

The Skelton Panel observed, "Innate talent probably is the most fundamental component for the development of a strategist."[334] However, it further declared, "Talent alone is insufficient; it must be reinforced by both appropriate experience and relevant education."[335] The Skelton Panel took a methodical view in maintaining that the PME system, as a whole, "should emphasize analysis, foster critical examination, encourage creativity, and provide a progressively broader educational experience with each level of schooling building on the previous level."[336] It asserted that all officers, whether or not they have the potential to think strategically or become strategists, would benefit from a broad three-tiered approach. At the first level, a strategist should develop "a firm grasp of an officer's own service, sister services, and joint commands."[337] At the second, he or she should develop a clear understanding of tactics and operational art – or knowledge of the employment of combat forces.[338] The Skelton Panel reasoned that lessons at these first two tiers could only be conveyed through PME.[339] At the third level, strategists should develop an "understanding of the relationship between the disciplines of history, international relations, political science, and economics," because each is "critical to the formulation of strategy."[340] As previously discussed, in promoting this final step, the Skelton Panel lauded the benefits of a supplemental civilian graduate education. It carefully noted: "Education outside the PME system may be necessary for the development of strategists, but it should not be viewed as a substitute for professional military education."[341] The services allow small numbers of officers to earn civilian degrees in strategy-related disciplines,[342] and constructive critics argue that the PME system is disproportionately devoted to achieving learning objectives at the first two levels (i.e., regarding service competencies, and the tactical and operational aspects of warfare) at noticeable expense to PME efforts at the third and culminating level.[343]

An Expanded Approach to Strategic Studies

The framework for teaching strategy at PME institutions has expanded since 1989. This expansion has resulted in a broader approach to teaching strategy to mid-grade officers. The Skelton Panel recommended that national security strategy be taught solely at National and ICAF and that national military strategy be taught at each of the senior service schools.[344] Current law and policy support both a wider and deeper distribution of strategic studies. Law and policy now require those senior PME schools accredited to teach JPME II (i.e., National, ICAF, each of the senior service schools, and the Joint Advanced Warfighting School) to teach national security strategy and national military strategy.[345] Current law also requires national military strategy to be introduced earlier, at ILE schools, as a function of JPME I.[346] Policy further permits those schools to offer introductory courses in national security strategy to reinforce perspectives on national military strategy, theater strategy, and operational planning.[347] A degree of standardization helps align various strategy-related curricula among the senior PME schools and foster continuity between the senior and intermediate levels of instruction. It does not appear to limit the flexibility afforded the services' PME institutions to emphasize air, land, sea, and space components of strategy and joint warfighting.[348]

Recently established joint content requirements and advanced operational learning centers also help to build greater familiarity with strategic concepts at the intermediate level. JPME must include, at a minimum, thorough coverage of: 1) national military strategy; 2) joint planning at all levels of war; 3) joint doctrine; 4) joint command and control; and, 5) joint force and joint requirements development.[349] Consequently, the OPMEP incorporates introductory treatments of theater strategy and planning, national military strategy, and national security strategy within the joint learning areas of JPME I offered to in-residence students at ILE institutions.[350] The JPME I programs offered to non-resident students by the senior service PME schools contain more extensive treatments of OPMEP-prescribed learning areas in national security strategy, national planning systems and processes, national military strategy and organization, theater strategy, and joint strategic leadership.[351] Each of the services also operates a highly selective one-year graduate degree program at the intermediate level.[352] The Army, Navy, and Air Force programs blend advanced study in operational arts with strategic studies to accelerate the involvement of a select few in planning on high-level staffs.[353] The services monitor graduates of these programs to support a small cadre of joint operational planners within the middle ranks of the officer corps.

Observation: The services have developed selective one-year graduate degree programs at the intermediate level. The Army, Navy, and Air Force programs combine advanced study in operational arts with strategic studies to facilitate planning on high-level staffs. The services monitor the graduates of these programs to support joint operational planning functions.

Strategic Studies at the Senior PME Schools

Despite these developments, strategy is still primarily taught at the senior PME schools to O-5s and O-6s.[354] The law requires that JPME II curricula cover national security strategy, theater strategy and campaigning, joint planning processes and systems, and integrated joint, interagency, and multinational capabilities, in addition to the subject matter areas covered by JPME I.[355] The OPMEP further requires that institutions offering JPME II credit address key concepts in joint warfare, theater strategy, and campaigning in a JIIM environment, the integration of JIIM capabilities, and joint strategic leadership.[356] The OPMEP also offers guidance for designating and apportioning appropriate emphases among the senior schools.[357] The senior service schools emphasize strategic leadership, national military strategy, and theater strategies, while National and ICAF focus on national security strategy (with the focus at ICAF centering on strategic resources).[358] The senior service schools are also at liberty to emphasize service-specific aspects of strategy. The Joint Advanced Warfighting School is a senior school within NDU offering a single-phase JPME curriculum for prospective JQOs who expect to be assigned to planning positions on the Joint Staff at a COCOM, or on joint task forces.[359] Its curriculum emphasizes the integrated strategic employment of joint forces through "exercises and case studies in a joint seminar environment."[360]

All of the senior PME schools teach strategy to some degree through detailed analyses of historical case studies. The case-study method was proven effective in exploring strategy as an evolving and adaptable discipline at the Naval War College under Admiral Stansfield Turner's leadership in the early 1970s. He reportedly stated: "Studying historical examples should enable us to view current issues and trends through a broader perspective of the basic elements of strategy. Approaching today's problems through a study of the past is one way to ensure that we do not become trapped within the limits of our own experience."[361] Professor Murray testified that the Naval War College's strategic studies program continues to be the "gold standard" by which other senior PME schools' efforts should be measured.[362] Other senior-level schools continue to innovate with respect to strategic studies. USAWC elevated its program in 1998 when it established the Advanced Strategic Art Program to broach difficult case studies with exceptional second-year students, and some ICAF faculty members have proposed establishing an Advanced School for Strategic Planning and Strategy, again for select second-year students, to examine national strategic resources and capabilities.[363]

PME for Flag Officers

The PME system offers little formal instruction to flag officers. The six-week Capstone course for newly selected general and flag officers offers a rare opportunity to enhance strategic thinking among senior officers. The Capstone course offers its participants JPME III credit and a timely orientation among the higher decision-making levels of the services, the COCOMs, the Joint Staff and other joint commands, other governmental organizations, and those of the United States' international partners. It also promotes the building of relationships among participants from each of the services and other governmental agencies, which flag officers may carry forward into positions of high authority. The Capstone course attempts to synthesize a familiarity with geostrategic concepts and the functional aspects of various strategic-level authorities.[364] Critics contest that the Capstone course's brief duration, its lack of accreditation, and its executive learning approach limit its value in developing strategic thinkers; that it does not meet the required reading or peer discussion standards of civilian executive learning models; and, that its requirements should be revisited.[365]

The Capstone course is currently undergoing an internal review. In a memorandum, dated 23 June 2009, the CJCS advised the NDU President that:

> CAPSTONE must maintain focus on resourcing and authorities at the strategic level without sacrificing the essentials of executive command. Prioritize engagement with principal leaders at the combatant commands and Services, with more time spent in command centers and joint intelligence operations centers, and fewer command overviews in conference rooms.[366]

Although this direction encourages interactions with strategic decision-makers, it does not prescribe significant change with respect to the Capstone course's treatment of strategy as a subject. The Subcommittee expects that a response to the CJCS from the NDU President will be forthcoming.

Further educational opportunities for flag officers are limited. The Service Chiefs are responsible for administering brief supplemental seminar programs for flag officers to provide "broad perspective of the operational and strategic levels of war."[367] The Combined/Joint Force Functional Component Commander Courses (air, land, and maritime) are one-week programs that prepare prospective theater-level combat leaders from every service at the one-, two-, and three-star levels. The Joint Flag Officer Warfighting Course is a two-week program for preparing potential theater combat commanders, COCOM service component commanders, and joint task force commanders. The Service Chiefs are also responsible for administering the Senior Joint Information Applications Course at their respective war colleges to reinforce flag officer leadership capacities with respect to information and cyberspace operations. The Combined/Joint Force Special Operations Component Commander Course is a more specialized one-week program for flag officers that is sponsored by the U.S. Special Operations Command. Each of these courses is limited to 18 attendees and is offered on a semi-annual basis. The services may also designate select one- and two-star officers who are "concerned with strategic planning and the economic, efficient, and effective allocation and use of scarce defense resources in today's complex and uncertain security environment" to attend the Senior International Defense Management Course.[368]

Educational opportunities for senior flag officers are also limited. The two-week Pinnacle course assists "prospective joint/combined force commanders" (i.e., three- and four-star flag officers) in developing an "understanding of national policy and objectives with attendant international applications and the ability to translate those objectives and policies into integrated campaign plans" for use in a "complex global environment."[369] The Army and the Navy also offer abbreviated executive education programs. These courses emphasize strategic business perspectives, force management, and international relations, especially for service-specific O-9 and O-10 billets.[370] Lieutenant General Barno contends that limited educational opportunities for flag officers may create an inverse relationship between intellectual development and strategic responsibility. As senior leaders ascend, they grow more distant from their educational groundings, and they must increasingly rely on experience, self-study, and personal relationships to develop their strategic acumen. This discontinuity may undermine efforts to meet demands for strategic leadership in a complex and uncertain security environment.[371]

Observation: PME is a factor in cultivating strategists, but it is not currently the primary means for shaping future strategic decision-makers. Outside educational opportunities, training, experience, and mentorship also contribute to joint and service-specific efforts for doing so.

Observation: The PME framework for teaching strategy has expanded since 1989. Law and policy require national security strategy to be taught at all senior PME schools offering JPME II credit. Strategy is still primarily taught at senior levels, but law and policy also require that the intermediate PME schools offer introductory coursework in national security strategy, national military strategy, and theater strategy and planning.

***Finding*: PME for flag officers is limited. The complex and rapidly changing security environment may require greater educational continuity for senior strategists.**

***Recommendation*: The CJCS and the Service Chiefs should evaluate whether additional or more rigorous requirements for educating senior strategists should be established.**

The requirement for military strategists is dependent on systematic identification and cultivation of rare talent. This requirement is even more pronounced by an increasingly complex and ever-changing security environment. There are numerous demands for strategic analysis at various levels of organization. However, the military should identify and cultivate the most creative and innovative of its strategic thinkers for the significant intellectual and leadership challenges of command decision-making at the strategic level. Joint and service efforts for developing strategists involve PME, outside educational opportunities, experience, training, and mentorship. The Army and the Navy have the most advanced systems for cultivating strategists, but each of the services should review and further develop their processes for identifying and cultivating strategists to optimize the development of qualified strategic decision-makers. Each should endeavor to balance academic achievement in a strategy-related discipline with actual command experience. PME is an integral factor in developing strategists, but it is not the exclusive, or primary, means for doing so; nor, should it be. Despite welcome efforts to broaden and deepen educational exposure to strategic studies within the PME system, PME's most significant contributions to the shaping of strategists are broadly offered to mid-grade officers at the senior PME schools. The CJCS and the Service Chiefs should revisit whether more rigorous education is warranted for senior leadership.

Delivery and Content

Apart from underscoring the specific need to identify and cultivate strategists, the Skelton Panel also evaluated the teaching practices—pedagogy—and rigor as well as the curricular content of PME courses. In addition to focusing on developing strategists, which the Panel viewed as a scarce national resource, the Panel made significant recommendations for raising educational standards at the schools that were more broadly aimed at improving PME for the entire officer corps. In light of the Panel's criteria for teaching excellence, this Subcommittee applied the Skelton norms to current practices at the schools to assess the degree to which they had made progress in advancing pedagogy and increasing rigor. Finally, the

Subcommittee looked generally at four areas of PME curricula to evaluate how the schools balance emerging and enduring subject matter.

Pedagogy

"How an institution teaches its curriculum can be as important as what is taught."[372]

The Skelton Report, 1989

The Skelton Panel defined pedagogy as "the art, science, and profession of teaching."[373] In examining pedagogy, the Panel primarily focused on what it described as "active versus passive learning," with a decided preference for the former.[374] Participating in small seminar discussion groups typified active learning, while passive methods included observing lectures, symposiums, panels, and films. The panel favored other pedagogical practices including student engagement in independent research projects and greater involvement in elective coursework. The increased use of simulations and war games, at that time, was viewed as a positive development.[375]

Active versus Passive Learning

In the Officer Professional Military Education Policy (OPMEP), the CJCS has articulated the intent and even the language of the Skelton Panel and its standard for evaluating instruction methods. Schools are assessed on the degree to which they "employ predominately active" instructional methods and active student learning.[376] The Skelton Panel did not explicitly identify a target ratio of active to passive hours. Nevertheless, it considered 49 percent passive hours at one school excessive, and it praised another for its "commendably low 10-percent passive education" and regarded this measure as a "goal for the other schools."[377]

The Officer Professional Military Education Policy (OPMEP), like the Skelton Panel, does not mandate a numerical standard for active learning.[378] This study did not undertake a detailed hour-by-hour analysis of instruction conducted through active learning. The PAJE teams do this routinely during their assessments of each of the 12 PME institutions that receive them, but they do not publish figures in every case.

Of the 12 most recent PAJE assessments, reports cited numerical data for four of the schools. In 2008 among the four, percentages for active learning, which included "student-centered seminar discussion, case study, simulation exercises, and field research," ranged from 90 percent for the Joint Advanced Warfighting School to 65 percent for the Air Command and Staff College (ACSC).[379] The latter case was the only one of the 12 PAJE reports that highlighted a shortcoming in this area. The PAJE report suggested the school should examine its JPME curriculum "in order to improve the active learning ratio."[380]

Observation: The PME schools promote more active student participation in the learning process. Active learning criteria are evaluated by the PAJE teams and all 12 PME institutions reportedly "employ predominately active" instructional methods.

Seminar Discussion Groups

Student-centered seminar discussions are the core means of instruction at the in-residence intermediate and senior PME institutions and exemplify the active approach to learning. Even though all 12 schools were found by the respective PAJE teams to meet the standard of "employing predominately active" instructional methods, these assessments were based primarily, if not solely, on a paper curricular review, so they are not necessarily qualitative measures of the active components of the curricula. In some instances, the PAJE reports do, however, comment on seminar size.

The ideal seminar size is an oft-discussed topic in the PME institutions. For example, the Subcommittee heard from the senior leaders at National that a seminar size of 13 students was considered ideal for pedagogical purposes and was the goal at that school.[381] *The Chicago Handbook for Teachers, A Practical Guide to the College Classroom* reached a similar conclusion, stating that seminar discussion "works best with a group of eight to fifteen."[382]

Recent initiatives for more international, interagency, and, in some cases, industry participation, are creating pressure to expand seminar size. At the service schools especially, there is tension between maintaining their distinctive character and facilitating joint acculturation by adding interagency and international students to the seminars. During its visits, the Subcommittee found that all of the schools divide students into seminar discussion groups of between 9 and 17 students.[383]

Although the OPMEP does not explicitly establish an ideal range, the most recent PAJE reports specifically mentioned seminar size in four cases. In the January 2009 evaluation of the Marine Corps War College, the PAJE study reported that the small seminar size of 9-10 students "lends itself to excellent student interaction."[384] The October 2007 assessment of Air Command and Staff College described the seminar size of 12-13 students as facilitating "a dynamic learning environment."[385] In the case of the College of Naval Warfare, its most recent PAJE evaluation observed that its "(s)eminar size of 14-15 students is well-designed for small group instruction and active learning methods."[386] In a disapproving note, the most recent PAJE for Army War College recognized that "the core curriculum emphasizes seminar discussions as an active learning method," however, it pointed out that "with 17 students per seminar, the College is operating above the optimal size (10-12 students) for Socratic seminar instruction."[387]

Another important pedagogical factor is student body composition within the seminars since the OPMEP establishes standards for class and seminar mix. Senior service schools have a cap of 60 percent for students representing the host service or department in the entire student body. The student body for this purpose includes international officers and civilians. Each seminar, according to the OPMEP, must have at least one student from each non-host service or department. While the intermediate-level service schools do not have a percentage limit for their student bodies, their seminars must also have at least one student from the non-host services or departments. In the case of National, ICAF, and the Joint Advanced Warfighting School, there must be "approximately equal representation from each of the three military departments in their military student bodies."[388] Student quotas for Joint and Combined Warfighting School are allocated on the basis of service representation on the JDAL.[389]

Recent PAJE reports reflect closer attention to student body composition by service than in the past. Since January 2009, all four reports made observations on non-host service participation, where previously, going back to 2003, there was only one mention. While all the schools met the OPMEP minimum requirements, it is likely that this signals concern that the strains of the current operational environment could affect the ability of the services to meet their commitments.

The January 2009 PAJE report for Marine Corps War College noted that in each of its 9-10 student seminars, there were two or three students from each non-host service and that each had one or two civilian students, which is approximately the same representation as at the joint schools.[390] In the subsequent two reports at the senior-level for the College of Naval Warfare and Army War College, there was specific notation that there was no more than 60 percent host military department representation in the student body.[391] For the Navy's intermediate-level College of Naval Command and Staff, the PAJE observed that there was at least one officer from the non-host departments in each seminar group.[392]

What the OPMEP does not specify and PAJE studies do not necessarily capture, however, is the degree to which the background and specialties of the students contribute qualitatively to an effective seminar. A Marine student, for example, meets the OPMEP requirement for a Navy Department student, but that student may not bring a breadth of perspective on the naval contribution to joint warfare. The Subcommittee heard from a Marine major at the Command and General Staff College that the students in his seminar expected him to provide expertise in general maritime matters and that he was challenged to do so. In other seminars, a non-host student might, by virtue of his specialty, have very little background in the overall or other specialized capabilities of his or her own service. The Subcommittee encountered medical corps officers and chaplains in seminars who ostensibly met the OPMEP requirement for service representation. They admitted that their ability to contribute to the seminar professionally was limited when engaged in operational planning.

Although the OPMEP is primarily concerned with joint (as opposed to interagency) acculturation in this regard, and does not set standards for numbers of interagency or other civilian students, there is an increasing demand by the service schools for more State Department and USAID students in particular. The Subcommittee also encountered, in some instances, State Department students who were not Foreign Service Officers, but were from that department's diplomatic security bureau. Although these students benefited from the in-residence PME, they were not usually able to provide the diplomatic or development insights that would be of optimal benefit to the seminar.

Finding**: While all the PME institutions meet the OPMEP numerical requirements for non- host service and departmental representation in seminar composition, the PAJE process does not evaluate the important qualitative impact that students and their career specialties have on seminar dynamics.**

The Case Study Method

The "case study method" is often broadly applied to the analysis of hypothetical or real-world issues, situations, and problems in which students place themselves in the roles of decision- makers. The well-known Harvard Business School case method is frequently cited as a model of this approach. Frequently, the case in question is illustrative of a cautionary or exemplary principle. This is sometimes confused with the "case method" used to study the law by examining judicial opinions. What the two methods often share, however, is the active involvement of students who are generally guided by a faculty member teaching through the use of hypotheticals and incisive questions.

The Skelton Panel did not precisely define the term, but it is clear from its numerous references what the Panel had in mind. In discussing the use of case studies in developing the ability to formulate strategy, the Skelton Report suggested presenting students with what it called “modern problems in strategic choice.”[393] In analyzing recent historical cases, students were to rely, to the maximum extent possible, on the:

> . . . original documents and evidence that were actually available to contemporary decision makers as they faced national security problems and tried to develop adequate responses, so that the real intellectual difficulties and limits facing the makers of strategies are recreated.[394]

Students may determine whether or not the policy makers, decisions were well-founded or whether alternatives should have been chosen. The Panel also saw a relevant application of this method for educating officers in the employment of multi-service forces, drawing on the precedent of its application for this purpose at the World War II-era Army-Navy School, where it was “designed to give the students an understanding of the capabilities and methods of each service.”[395] The Skelton Panel endorsed the case study methodology because it saw value in the careful study of military history, the classroom interaction inherent in thoughtfully discussing both lessons learned from past problems and their application to current issues, and the development of creative solutions to joint warfighting challenges.

The schools’ curricula include, and the PAJE reports confirm, an appreciation for the value of case studies in coursework. The three most recent reports make special note, citing the use of case studies in support of the blocks of instruction. At the College of Naval Warfare the PAJE report observed, among other numerous instances, that the “[c]ase studies in the [Strategy and Policy] course provide linkage between strategy and policy development and the national and joint planning systems and processes across a wide range of joint military operations.”[396]

Rigor

> *“The question of rigor is best answered individually, especially when you are at war.”*[397]
>
> **General Martin Dempsey, USA Commanding General,**
> **U.S. Army Training and Education Command**

Although the Skelton Panel defined academic rigor in terms of a threefold combination of a challenging curriculum, established standards of performance, and student accountability to those standards, it focused primarily on the aspect of student accountability. Specifically, it devoted considerable discussion to the question of whether students should be tested and graded on their coursework.[398]Although the Panel reported that all of the intermediate-level schools at the time had graded exams, the same was not true for their senior-level counterparts. The Panel singled out the College of Naval Warfare as the only senior school to administer them. The Panel unambiguously noted that none of the other senior schools administered tests whatsoever. While the Panel reported that some of the intermediate schools provided examination scores solely for student reference, it did not describe the information that appeared on the student transcripts, which renders it difficult to compare current grading systems with the Panel’s recommendations.[399]

In observing these varying student evaluation practices at the time, the Panel expressed some decidedly firm views on the subjects of frequent essay examinations and graded coursework. While acknowledging that students, especially those at the senior level, were self-motivated individuals who were already the product of a rigorous selection process, the

Panel did not consider this a compelling rationale for not grading student work. The Panel adamantly argued that "although an individual student may impose rigorous standards on himself regardless of a school requirement, the sine qua non of a PME school's rigor is graded activities."[400]

Nevertheless, considerable leeway was left to the schools. The Panel did not mandate letter grades corresponding to a numerical range, e.g., an "A" for 90-100 percent, or even letter grades based on a subjective judgment. It characterized the Marine Command and Staff College's (Marine C&SC's) system at the time of using the grading terms "non-mastery," "mastery," and "high mastery" as "intriguing."[401] The Panel was silent on whether a "pass/fail" was sufficiently stringent.

Like the Skelton Panel, the Subcommittee observed a wide variety of practices for student assessment across the 12 senior and intermediate in-residence PME institutions. All of the intermediate and senior schools now have graded examinations and some form of final grades. A number of institutions record letter grades while others use a pass/fail system. In the latter case, the absence of letter grades on official transcripts does not necessarily mean that the faculty does not provide students with assessment of their performance. While an official transcript may only record a grade of "pass," in most cases the students receive a more comprehensive evaluation in the form of directed specific comments on their work.

The USAWC, for example, only enters pass/fail marks on student transcripts. Students receive numerical scores of 1 through 5 on their work, however, in accordance with the degree to which they achieve course learning objectives.[402] On the other hand, the College of Naval Warfare and the Navy's CNCS award letter grades with plusses and minuses for core courses based on a numeric score.[403] Electives are graded on a three-tier scale of high pass/pass/fail.[404] These grades are recorded on the official transcripts.[405]

Even within NDU there are differing approaches. National uses a similar system to that of the USAWC. All courses are graded on a pass/fail basis.[406] Similarly, the faculty at National augments this with ratings for student use on the constituent requirements, in this case whether student performance is "above standards," "meets standards," or "below standards," along with narrative evaluations in a number of categories for each core course, elective, paper, presentation, and oral exam. Within the same university, ICAF assigns letter grades with plusses and minuses on overall course evaluations.[407] The Subcommittee heard from some students at National that the faculty appraisals of written assignments and examinations were generally more useful and reflected more demanding standards than letter grades lacking a more comprehensive critique.[408] Dr. Breslin- Smith, a retired professor and former department head from National, spoke highly of what she called the "challenge of the scenario analysis" posed by her former college's two annual oral examinations. Dr. Breslin-Smith testified: "If a student can analyze the components of a given scenario, its strategic implications, and thoroughly respond to the in-depth questions prompted by the discipline of the framework, we can assess the rigor of the student's thought and preparation."[409]

Overall, the PME institutions have addressed the Skelton Panel recommendations for raising academic standards. At the Subcommittee's final PME hearing, Dr. Williamson Murray commented that, although the Naval War College still maintains its position as the leader in this regard, "the improvement in academic rigor in the staff and war colleges has been considerable since the late 1980s."[410] Professor John Williams cautioned against much conformity when measuring student performance. He supports "a great deal of discretion to the educators and administrators at the various PME institutions, subject to a common understanding on the importance of academic rigor."[411]

***Finding*: PME institutions have generally implemented the Skelton Panel recommendations for more demanding standards. All of the schools have writing requirements, and all of the senior-level schools now have graded exams. Although PME institutions have retained or adopted a variety of approaches with regard to grading, the Subcommittee found that pass/fail systems based upon objective learning standards and supported by comprehensive and timely feedback, even if provided only for student use, do not necessarily detract from the rigor of the academic programs.**

The OPMEP also speaks to the grading process itself, asserting that the PME institutions' "(e)ducational goals and objectives should be clearly stated and that students' performance should be measured against defined standards."[412] Accordingly, all of the PME institutions have published grading standards that are reviewed by the PAJE process. The Army CGSC, however, has developed detailed grading templates, referred to as "rubrics," for each academic lesson.[413]

While the most recent PAJE report viewed the rubrics favorably for creating clear expectations for performance, a number of students did not share this perspective.[414] The Subcommittee heard comments that some faculty applied the rubric to writing assignments in a mechanistic fashion, promoting conformity to set criteria and discouraging more complex or creative approaches to the topic. One student spoke of having to "dumb down" his writing to fit the rubric which he referred to as a "checklist."[415] One faculty member remarked that "the rubric tells you what to think, not how to think. This makes it easier for the faculty to grade, but it's not good for the students."[416]

Observation: Scoring tools such as the Army CGSC rubric allow for a standardized evaluation according to specified criteria, making grading simpler and more transparent. Caution is warranted, however, if rubrics are applied in a manner that would limit original or more creative approaches by students to the subject matter.

The Subcommittee also came across a number of issues related to academic rigor that the Skelton Panel did not necessarily assess. One relatively recent trend encountered was the move of PME institutions to award master's degrees and the attendant civilian accreditation process. The Subcommittee considered the introduction of the added scrutiny of the civilian accreditation process to positively influence the rigor associated with those PME institutions awarding advanced degrees. The Subcommittee did hear some views, however, that the requirements for master's degrees were a distraction from attention to PME.

During visits to the PME institutions, the Subcommittee heard a number of views on the effect of master's degree programs on the rest of the curricula. One student at ACSC spoke of "academic mission creep" resulting from the master's program, implying that the additional requirements tip the focus toward academic instead of professional education.[417] Additionally, the Subcommittee heard from a student at the Army CGSC that the supplemental work, centered around a thesis paper, for earning the Masters of Military Arts and Science (MMAS) degree, proved onerous for many students.[418] Although optional, many students viewed the MMAS program as the only opportunity that they were likely to have to earn a master's degree in the foreseeable future, especially under the current operational tempo. Moreover, an advanced degree was, at least in the students' minds, a positive discriminator for future promotion or a second career. One faculty member at that institution referred to the regular curriculum as the "bill payer" for the master's program, observing that the MMAS candidates tended to prioritize their efforts toward the thesis at the expense of preparation for daily classes.[419] The commandant of the USAWC noted that some of his students, when confronted with more rigorous academic challenges, found it "convenient to blame the master's degree and academics."[420] He went on to say that "while professional topics, demands and standards have increased, no new major, purely academic requirements have been added."[421]

When the leaders of the senior schools were asked if their master's degree programs diminished the PME mission, most viewed them as enhancing and reinforcing it, rather than detracting from it. More than one school leader mentioned that offering the opportunity to teach at an accredited graduate-degree program helped the schools attract higher-quality faculty members thereby improving the PME curricula and quality of teaching. Assuming that the PME schools recruit from the same limited pool of candidates with expertise and credentials in national security- related fields, it is conceivable that the schools with master's programs would be more attractive. Some school leaders also cited the positive impact that the accreditation process had in importing best practices to the military schools, especially in the areas of course development and curricular review. The commandant of National was the only leader to take an agnostic position on the value of civilian accreditation of graduate degrees, disavowing any increase in rigor as a result.[422]

Observation: Although the Subcommittee heard some concerns voiced over master's degree programs detracting from the PME mission, there was no clear evidence to this effect. There was general agreement among school leaders that the civilian accreditation process has a positive influence on rigor and that schools' graduate-degree granting status has a beneficial effect on faculty recruiting.

The Subcommittee did not originally intend to devote a great degree of attention to the Capstone course for newly selected flag officers, but its research coincided with a re-examination by the Joint Staff and the NDU president into the program's rigor. The Skelton Panel had previously identified a number of shortcomings with the Capstone course including a lack of a full-time faculty and what the panel considered an insufficient six-week course length.[423] Additionally, the Report assessed that the course failed to live up to its potential, in large part due to the lack of substantive academic work in joint matters and strategy. While it made no recommendation for graded exams or writing assignments, the Panel proposed an overhaul of the program and recommended that the "(c)ourses should be as rigorous and demanding as the students' future responsibilities will be."[424]

A previous version of the OPMEP incorporated learning objectives in line with the Skelton Panel's recommendations for the Capstone curriculum in the areas of national security strategy and joint operational art.[425] It was not until 2009, however, that the OPMEP added a PAJE requirement for Capstone.[426] This formal requirement was preceded by a PAJE-like independent review for the CJCS.[427] The review concluded that the Capstone course was "generally meeting the established requirements for joint service education" and proposed no marked changes to the curriculum. At the same time, it found that there was no effective tool or method to determine how well the newly selected flag officers were mastering the learning objectives.[428]

The CJCS recently instructed the NDU president to develop a "course assessment mechanism" to evaluate the effectiveness of the instruction. Discussions on the best way to achieve this revealed a tension between competing educational models: the executive learning model, which assumes a large measure of self-motivation on the part of the adult students, and a program with demonstrable educational objectives. At the time of this report, an internal review was still being conducted, although the forthcoming evaluation mechanism may involve measurable performance in a short (1-2 day) classroom exercise[429] and possibly writing requirements on strategic perspectives gained from the U.S. and overseas trips.[430]

Observation: There is tension with the Capstone course between the OPMEP's academic model with its PAJE accreditation requirement, and with the current executive education model. The CJCS has given the NDU President direction to revise the course.

Curricula

> *"The panel believes that the primary subject matter for PME schools and, consequently, the underlying theme of the PME framework, should be the employment of combat forces, the conduct of war."*[431]
>
> **The Skelton Report, 1989**

In examining curricula, the Skelton Panel concentrated mainly on the extent to which the service PME institutions were meeting the requirement, new at the time, of integrating joint content into their coursework required by the Goldwater-Nichols reforms. The Panel conducted a detailed review of each school's attention to joint subject matter and went as far as determining the percentage of each school's curriculum that it considered "joint." While the Report confirmed the services were devoting a sufficient quantity of their core curricula hours to joint subjects, it concluded that the "(d)iscussion of joint material focused almost entirely on the role of the parent service in the joint operation or activity."[432] It recommended a more "well rounded approach to joint education in the service schools."[433]

In this study, it was not the Subcommittee's intent to replicate the hour-by-hour curriculum analysis of the Skelton Panel. The Military Education Coordination Council (MECC) has regularly performed that accounting for each of the PME institutions as part of the PAJE process. The MECC systematically assesses the curricula of the schools looking at the extent to which they "prepare graduates to operate in a joint, interagency, intergovernmental, and multi-national environment and bring a joint perspective to bear in their tactical, operational, strategic, and critical thinking as well as professional actions."[434]

The Subcommittee endeavored to evaluate how the PME institutions balance the traditional or enduring subjects in their curricula with new content and thinking emerging in response to the changing security environment, lessons learned from current conflicts, and what some see as anticipating the changing character of war.[435] They each have to decide, for instance, whether to change emphasis from teaching traditional texts like those by Sun Tzu and Clausewitz to covering the works of the experts on topics such as terrorism, counterinsurgency, and cyber warfare.

Apart from the constraints of covering the subject matter required by the OPMEP, the schools are given considerable latitude in deciding how to manage their curricula. Each service and its PME institutions formally review their curricula on a regular basis, to incorporate any new guidance as well as lessons learned from current operations, strategies, and policies. The Marine Corps and Air Force have processes that are illustrative of those in the other services. As introduced previously, the Marine Corps has a Curriculum Review Board at Marine Corps University which reviews the service's PME curricula every two years and makes recommendations for substantive changes based on service needs, Joint Staff, and combatant command input. On an annual basis, the individual Marine Corps schools conduct Course Content Review Boards evaluating each block of instruction incorporating instructor and student feedback.[436] To assist the schools in mediating with senior commanders making recommendations for new curricular matter, the Air Force, has established the Air Force Learning Committee to evaluate these kinds of suggestions.[437]

The Subcommittee examined four curricular areas to illustrate the balancing process: irregular warfare and stability operations; language, culture, and regional expertise; joint, interagency, intergovernmental, and multinational operations; and history. The first three deal with newer subject matter or that which is of growing importance in the current security environment. The last was chosen because history has traditionally formed the foundation of PME curricula, and the Subcommittee heard from faculty on more than one occasion that a reduction in history was being used as the "bill payer" for the addition of new subject matter into the curricula.[438]

In their testimony before the Subcommittee, school leaders often spoke of this challenge. The deputy commandant of the Army CGSC alluded to this dynamic, "We realize that friction will always exist where the current curriculum competes with the many emerging topics."[439] There is constant pressure from field commanders, according to the commandant of National, to include tactical and operational lessons learned. He underscored the need to protect what he called the "core elements" in the school's educational requirements and keep the focus of the curriculum centered on the strategic level of warfare.[440]

Irregular Warfare and Stability Operations

The Skelton Panel now appears prescient in its discussion of military participation in strategic thinking. In 1989, it cited critics of U.S. strategy as having "[t]he concern that U.S. military capabilities are inappropriately skewed toward unlikely contingencies and as a result, are inadequate for more probable low-intensity conflict."[441] That is not to say, however, that the lessons of guerrilla warfare and counterinsurgency had completely disappeared from the PME curricula.

Even before the current conflicts in Iraq and Afghanistan, the PME institutions were, to some extent, teaching irregular warfare (IW), using case-study examples such as occurred in the American Revolution, Algeria's war for independence, America's involvement in the Vietnam War, and others.[442] Irregular warfare, a term that has replaced "low-intensity conflict," involves insurgency, counterinsurgency, terrorism, and counterterrorism,[443] and refers to the:

> [V]iolent struggle among state and non-state actors for legitimacy and influence over the relevant population(s). Irregular warfare favors indirect and asymmetric approaches, though it may employ the full range of military and other capacities, in order to erode an adversary's power, influence, and will.[444]

A more serious consideration of irregular warfare has emerged as a result of the military's experience in Iraq and Afghanistan. The 2008 *National Defense Strategy* identified improving proficiency in irregular warfare as the Department's "top priority." In discussing "modes of warfare," the document calls for a displaying "a mastery of irregular warfare comparable to that which we possess in conventional combat."[445] Given this degree of prominence, one might expect to see a corresponding shift in the PME system and its institutions' core curricula and electives.

The most recent version of the OPMEP adopts this new emphasis and directs the intermediate and senior PME institutions to devote attention to irregular warfare. The Chairman's Vision section of the instruction has been updated, adding language on improving the military's capability to "wage, as necessary, traditional *and irregular warfare*."[446] Moreover, the OPMEP acknowledges the equal importance of the two forms of conflict.[447]

The associated change to the Joint Learning Areas and Objectives section of the OPMEP,[448] however, does not echo the "top priority" urgency expressed in the high-level policy guidance. It was amended with a new requirement for students to simply "(c)omprehend the fundamentals of traditional and irregular warfare" at both the intermediate and senior level colleges.[449] While the language in the requirement does not ignore or marginalize irregular warfare, asking students to merely comprehend "fundamentals" does not seem to fully capture the force of the *National Defense Strategy* or even the Chairman's Vision, for that matter.

Observation: The Joint Learning Areas and Objectives section of the 2009 OPMEP does not impart the same level of importance to irregular warfare as the National Security Strategy and Chairman's Vision statement on PME.

As the PAJE cycle is just beginning to assess the schools' responses to the new guidance, it is too early to determine, in a systematic way, whether irregular warfare is being treated on par with traditional warfare. Still, the Subcommittee received enough testimony from the school leaders to make some preliminary observations.

School leaders offered the Subcommittee their descriptions of core courses with irregular warfare content, electives specifically devoted to irregular warfare (IW), and other relevant activities, such as special seminars. Illustrating the rationale for the integration of IW-related topics and other subjects into the core courses at the Army CGSC, its Deputy Commandant pointed out that the curricula there "is not organized into discrete blocks such as individual classes in counterinsurgency, stability operations, threats, culture, major combat operations, etc."[450] Instead, an "integrated curriculum approach," he explained, "...allows [the CGSC] to address multiple learning outcomes in common lesson blocks."[451] The Director of the Marine C&SC highlighted the school's electives which are representative of the other career-level schools. Those directly related to irregular warfare include:[452]

Insurgency from the Insurgent Perspective
Airpower and Asymmetrical Warfare
The American Indian Wars: Irregular Warfare Relevant to the 21st Century
Counterinsurgency Theory and Practice
The Vietnam War

The same year that the *National Defense Strategy* elevated irregular warfare as the top priority, the Naval War College established a Center on Irregular Warfare and Armed Groups which hosts conferences and promotes research and teaching on this field.[453]

Observation: Although the Military Education Coordination Council has only recently begun assessing the incorporation of irregular warfare into the curricula of the PME institutions, the schools' began integrating irregular warfare subject matter into core courses and offering electives directly related to irregular warfare prior to the formal requirement in the 2009 OPMEP.

Observation: Given the PME institutions' pre-existing irregular warfare treatment, most schools will likely satisfy the new Joint Learning Area and Objectives within the OPMEP with their existing course offerings.

Stability operations are often associated with counterinsurgency, and, at times, with irregular warfare, but they can also be conducted apart from irregular warfare in cases such as foreign disaster relief or foreign humanitarian assistance.[454] Stability operations may also support major conventional operations, especially during the last two phases of the military's six-phase model, Phase V (Stabilize) and Phase VI (Enable Civilian Authority).[455]

The Subcommittee began an examination of stability operations in September 2007 from the perspective of Provincial Reconstruction Team (PRT) experiences in Iraq and Afghanistan. The Subcommittee also treated this examination as a case study in the larger context of interagency cooperation and a "whole-of-government" approach to complex security challenges.[456] While it did not specifically look at educational preparation in the PME institutions for military service in stability operations, the Subcommittee did recognize that the mission is likely to continue into the future, and it identified the need to develop an enduring capability for PRT-like missions.[457]

U.S. Army Captain, assigned to the 414th Civil Affairs Battalion, goes door to door to meet the women of Assyria, Iraq, in 2006, and invite them to the new women's center on the day of its grand opening.

Originally defined simply as "[m]ilitary and civilian activities conducted across the spectrum from peace to conflict to establish or maintain order in States and regions," stability operations now encompass "various military missions, tasks, and activities conducted outside of the United States in coordination with other instruments of national power to maintain or re-establish a safe and secure environment, provide essential governmental services, emergency infrastructure reconstruction, and humanitarian relief."[458] The definition clarifies that stability operations occur on foreign soil and not domestically. In addition to work by PRTs and Human Terrain Teams deployed to Iraq and Afghanistan, stability operations also include relief efforts like those recently conducted in Haiti in response to natural disaster.

As part of "rebalancing" the armed forces to be able to address a wider range of contingencies, the 2010 Quadrennial Defense Report (QDR) calls for increasing the capacity of the general-purpose forces to conduct, among several other tasks, stability operations.[459] The significance of stability operations is not only reflected in programmatic decisions, such as the Army procuring more Stryker vehicles and the Navy bolstering its riverine craft fleet, but it adds a dimension to policy and doctrine as well.[460] In a significant shift, the Department promulgated a policy in 2005, which was reissued in 2009, elevating stability operations to the level of a "core U.S. military mission that the Department of Defense shall be prepared to conduct with proficiency equivalent to combat operations."[461] In its discussion of developing future military leaders, the QDR specifically calls for continuing to place special emphasis on stability operations in PME.[462]

In light of this increased emphasis, one could reasonably expect to see more developed studies of stability operations enhancing the curricula of the PME institutions, including in simulation exercises. One might also expect to see the mandated Learning Areas in the OPMEP revised to incorporate this change. The OPMEP does not include stability operations as a learning area emphasis at the operational level. Rather, stability operations material is obscured as a component of "stability, security, transition, and reconstruction" in an enumeration of functional warfare specialties to be covered such as logistics, intelligence, and strategic communications.[463]

For this reason, it is difficult to gauge the attention the PME institutions pay to stability operations on the basis of the two most recent PAJE reports, which were completed following the publication of the current OPMEP. The September 2009 PAJE report for the USAWC did not utilize the July 2009 version of the OPMEP but instead that published in 2005 which does not mention stability operations.[464] The October 2009 PAJE study of the Navy's CNCS used the Learning Areas in the new OPMEP, commenting on the incorporation of cyberspace operations and weapons of mass destruction effects, but there is no reference to the integration of stability operations into the curriculum.[465]

***Finding*: The OPMEP has no distinct Learning Area for stability operations, despite those operations being recognized as a core military mission comparable to combat operations since 1995 by Departmental policy, which directed that stability operations be "explicitly addressed and integrated across all DOD activities," including those involved in education.**[466]

***Recommendation*: The Director of the Joint Staff should review the OPMEP to determine whether it adequately conforms to DOD Directive 3000.05, *Stability Operations*. If it does not, the OPMEP should be revised.**

The Subcommittee asked intermediate PME school leaders to describe the extent their simulation exercises and war games incorporated stability operations concepts. They all reported progress in this area and gave detailed accounts of their programs. If there was a common trend, it was the introduction of stability operations as a factor in scenarios involving adversaries employing traditional methods of warfare. For example, the commandant of the JFSC described exercises that included a five-day event at Joint and Combined Warfighting School which, although based on a traditional warfare scenario, called for the writing of a detailed operations plan with Phase V (Stabilize) and Phase VI (Enable Civil Authority) components.[467]

Most of the schools also run exercises exclusively devoted to stability operations. The Deputy Commandant of the Army CGSC, for instance, cited a 24-hour exercise at his school that focused on planning for a post-conflict scenario "where there are remnants of conventional forces operating as well as insurgent activities and large displaced populations."[468] Students are required to develop a "whole-of-government" approach with plans for working with local authorities "to restore rule of law, provide essential services and train and prepare host nation forces to work without significant outside support."[469] The president of the Naval War College described an exercise at the College of Naval Command and Staff that centered on providing humanitarian assistance in North Korea after a severe famine caused the breakdown of order and government capacity there.[470]

The "Nine Innings" exercise at the Marine C&SC appears to take a somewhat original approach. The task for students is to develop a four-year "Phase 0" or pre-conflict interagency plan that is designed to prevent future instability in a country with potential for disintegration, such as the Philippines. Additionally, instead of using a prepared scenario or classified information, the exercise draws completely on open source material related to the existing conditions and embraces any real- world events occurring in the country during the period and the exercise immediately leading up to it.[471] On the whole, the schools appear to be ahead of the formal requirements in incorporating stability operations into their curricula. While the PME institutions may be adapting readily to these new demands, their response has been in reaction to, rather in anticipation of, the new security environment.

Observation: Despite the OPMEP lacking any distinct learning area for stability operations, the services and joint schools are incorporating stability operations into their simulations and planning exercises at an appropriate level of emphasis, consistent with their status as core military missions.

Language, Culture, and Regional Expertise

Although they are not new problems, the conflicts in Iraq and Afghanistan have underscored the importance of foreign language and cultural competency for the armed forces. Only in recent years has the Department explicitly identified foreign language skills and regional expertise as "critical warfighting skills."[472] In its 2008 report on the development of language skills and cultural competency in the military, the Subcommittee found that despite departmental aspirational goals for creating foundational language skills for the general-purpose forces, the services were focused more on developing a culturally aware force than a linguistically capable one.[473] The PME institutions' efforts are similarly targeted. This, in part, reflects the lack of a requirement for language study in the OPMEP.[474]

The 2010 Quadrennial Defense Review (QDR) places equal if not greater weight on foreign language skills than on regional expertise and cultural skills. Under the major mission area of "Succeed in Counterinsurgency, Stability, and Counterterrorism Operations," the QDR calls for building expertise in the three areas of foreign language, regional expertise, and cultural skills, however, improving the foreign language capacity of the force receives particular emphasis. In discussing what key enabling capabilities U.S. forces need to perform more effectively, foreign language skills are singled out after improvements in rotary-wing aircraft, unmanned aircraft systems, and intelligence analysis.[475]

The PME institutions take varying approaches with regard to language instruction. Foreign language instruction is concentrated at the intermediate level, although there is some study in the war colleges. Only the Air Force and the Marine Corps have foreign language requirements for graduation. All of the schools struggle with the important demand for language expertise in irregular warfare. Their efforts, however, must grapple with the reality that measurable proficiency in a new foreign language, especially those commonly used in Iraq and Afghanistan, is difficult to achieve in 10 months of study dedicated totally to language, much less as a single component of a larger curriculum and only 30 to 50 hours of classroom instruction, at most.

Since academic year 2007-2008, the Air Command and Staff College has had a mandatory foreign language requirement with familiarization or "enhancement" as the goal. Until the current academic year, the school primarily used the *Rosetta Stone* software augmented by periodic assistance from visiting Defense Language Institute (DLI) instructors. Based upon student feedback, the school has shifted to 30 hours of classroom time with DLI instructors and has placed the language software module in a supplemental role. Students can choose from Spanish, French, Mandarin Chinese, or Arabic.[476] At Air War College, students take a non-credit elective in one of five languages (French, Spanish, Russian, Mandarin Chinese, or Arabic), also taught by DLI instructors.[477] While the Air Force has observed that due to limited exposure, it is very difficult for students to gain any more than a very basic proficiency, in the school's view, it still sends the signal to the students that this is an important skill.[478]

Like the Air Force, the Marine Corps has been adjusting its approach to foreign language teaching. Beginning in 2004, the Marine C&SC moved toward an ambitious 120-hour program in Modern Standard (formal) Arabic taught by contract instructors. Based on student and faculty feedback, the school transitioned in academic year 2008-2009 to 47 hours in the classroom with DLI instructors in Arabic, French, Korean, or Chinese, followed by a negotiation exercise in the target language.[479] In describing the exercise, the director of the school explained that "students must communicate in a rudimentary fashion to a non-English speaker in the chosen language and eventually turn over the negotiation to an interpreter."[480] Again, in response to student feedback and the changing national security environment, the school is replacing Korean with one of the languages spoken in Afghanistan.[481]

The Naval War College does not offer any language courses at either the College of Naval Command Staff or the College of Naval Warfare.[482] The rationale behind this decision, according to the school's leadership, is based on the amount of time required to achieve even a basic familiarity with a foreign language, especially those spoken in the current theater of operations.[483]

The Command and General Staff College has plotted a middle course, offering elective language classes in Arabic, Chinese, and French. After a week of what the school's deputy commandant described as "intensive classes" with DLI instructors, the students spend five months in an on-line program where they communicate with their DLI instructors by video teleconference. At the conclusion of the distance learning period, the DLI instructors return to the school for a final month with the students.[484]

Observation: Given the difficulty in imparting more than a rudimentary knowledge of a foreign language to students in 10 months of classes, the PME institutions have taken varying approaches to foreign language instruction as they balance the importance of language skills in irregular warfare and stability operations, where the objective is often the population, with the curricular trade-offs that have to be made within the core curriculum.

To a more consistent degree than language studies, the services and their PME institutions are focused on developing culturally aware forces. The Joint Learning Areas and Objectives in both the current OPMEP and its predecessor contain requirements to address "society, religion, and culture" at both the intermediate-level and senior-level institutions.[485] It is not unusual for the PAJE reports to make comments on the cultural and regional material in the curricula. However, the observations are selective, rendering it difficult to assess from them; it is necessary to rely heavily on the statements and testimony of the school's leadership. Although the schools' curricula were available for Subcommittee review, it was difficult to discern whether cultural material has simply been "relabeled" in the existing core courses or whether it has been more thoughtfully integrated into the subject matter.

All but one of the six war college-level institutions have a core course or a mandatory elective covering regional and cultural studies. National, ICAF, the Air War College, and the Marine Corps War College all address this subject area in a dedicated core course. They are variously named "Regional Security Studies," "Regional and Cultural Studies," or simply "Regional Studies." National's students are all required to take one regional studies elective related to their travel, as discussed below. None of the six core courses at USAWC are specifically devoted to this discipline, but students there must take one of their five electives in regional studies.[486] The College of Naval Warfare has neither a core course nor a required elective in regional studies but offers five regional area studies electives. While not mandatory, the school reports that a significant number of students per year enroll in these electives.[487]

There are international travel programs associated with regional study courses at the Air War College, National, and the Marine Corps War College. Although ICAF's Regional Security Studies course does not have a travel component, students can concentrate on a region and usually visit it in the international field studies trip connected with their Industry Studies core course.[488] Students at the Marine Corps War College are able to travel to several strategic regions in part because of their school's small size.[489]

At the intermediate level, the Marine Corps is the only service with a core course with the express purpose of addressing regional and cultural studies. According to the school catalog, its Culture and Interagency Operations course "is designed to improve students' abilities to understand and analyze regional cultures and the interagency components of national and international governments at the operational level of war."[490] Students at the Army CGSC, like their war college counterparts, are required to take one regional studies elective.[491] The Navy's CNCS does not have a required regional studies course in the core curriculum or mandatory elective, but the school does assign each seminar group one of five geographic combatant command areas of operations in the National Security and Decision Making course.[492]

This is not to say that the other schools' curricula are devoid of content in regional and cultural studies. There is an inherent cultural component in schools with language studies such as at ACSC. Additionally, there is a regional studies element intrinsic to the simulation exercises at all the schools which are routinely set in different parts of the world.

Joint, Interagency, Intergovernmental, and Multinational Operations

In the past decade, curricular standards in succeeding versions of the OPMEP have moved beyond "joint," in the limited sense of multiservice operations. They now encompass the JIIM aspects of operating in the present security environment.[493] The development of, and increase in, curricular standards for this subject area has been a feature of succeeding versions of the OPMEP.

The 2000 version of the OPMEP made no mention of JIIM.[494] By comparison, the 2005 OPMEP devoted a single Learning Objective to "joint, unified, and multinational campaign and operations" within the Learning Area of "Theater Strategy and Campaigning" to be taught at senior schools. The 2009 OPMEP, however, added two new Learning Areas: "Joint Warfare, Theater Strategy, and Campaigning in a Joint, Interagency, Intergovernmental, and Multinational Environment" and "Integration of Joint, Interagency, Intergovernmental, and Multinational Capabilities." These two Learning Areas now include 10 separate Learning Objectives. As the PAJE has not yet evaluated a senior school pursuant to the 2009 OPMEP requirements, it is too early to assess how well the senior schools are implementing these new curricular standards.[495]

For the intermediate schools, the 2005 version of the OPMEP devotes a Learning Area consisting of six Learning Objectives pertaining to "Joint and Multinational Forces at the Operational Level of War," but these objectives did not contemplate the interagency or intergovernmental aspects.[496] The title of this Learning Area remains unchanged in the 2009 version, but the Learning Objectives reflect greater attention to, among other factors, society, culture, and religion.[497] Although there are fewer curricular standards in the OPMEP for intermediate schools than for their senior counterparts, this may be appropriate as the services each approach JIIM considerations differently at the tactical and operational levels based on their service- specific needs.

If there are future candidates for inclusion to add to the holistic approach reflected in the JIIM acronym, both industry and non-governmental organizations (NGOs) merit consideration. In his 2010 written testimony for Congress, the commander of U.S. European Command underscored the necessity of military-private sector cooperation in countering piracy.[498] In his previous role as commander of U.S. Southern Command, he established a staff section for public-private cooperation. U.S. Southern Command established a number of initiatives, including goodwill activities on the part of NGOs and industry in support of U.S. engagement in the region.[499]

In his 2010 posture statement, the commander of the U.S. Pacific Command recognized the challenge of information sharing outside the limits of its subordinate military elements and higher headquarters. He stated that "[t]he ability to exchange information among DoD components, all levels of the U.S. Government, coalition partners, *and the private sector* is becoming increasingly important to regional operations."[500] Additionally, he highlighted efforts to create information technology infrastructure that would allow the command to communicate with industry partners.[501] Although private-sector collaboration is becoming progressively more important to combatant commanders, practice may not be at the point yet where it can be institutionalized in the PME system through adoption of dedicated OPMEP

Learning Area(s) and/or Learning Objective(s) which would cover cooperative operations with private entities.

U.S. Marine Corps Sergeant and Ugandan Captains, troubleshoot a tactical chat program that allows radio operators to send data over high-frequency radios.

***Finding*: As interagency and intergovernmental dimensions increasingly factor in the different levels of war, the Military Education Coordinating Council has added curricular standards related to joint, international, intergovernmental, and multinational (JIIM) considerations in joint operations. Since each service approaches JIIM differently based on how they operate within the joint environment, there may only be a need for a base line curricular standard.**

***Recommendation*: The Joint Staff should continue incorporating a base curricular standard with respect to JIIM and at the same time allow flexibility for the services to tailor the instruction best suited to their requirements.**

***Finding*: There is a necessity for combatant commanders to cooperate with the private sector in some operations such as combating piracy. There is potential for the combatant commands to engage in private sector partnerships with NGOs and industry to** support U.S. engagement activities.

***Recommendation*: The Joint Staff should consider incorporating OPMEP Learning Area(s) and/or Learning Objective(s), regarding cooperative operations with private entities**.

History

As mentioned previously, the study of history, and military history in particular, has traditionally formed the basis of the curricula at PME institutions. The Skelton Report's discussion of the importance of teaching history and its relationship to developing officers with an understanding of strategy, is worth quoting in its entirety:

> History, or more specifically the lessons of history, provides insights into how nations have adapted their military and security strategies over time to deal with changing

> domestic and international environments. Strategy is, after all, dynamic. It must take into account changing realities and circumstances. Military history is especially important. The history of combat operations, including an understanding of why a commander chose a given alternative, is at the heart of an education in strategy.[502]

This is not to say that military history should dominate the core curricula, but that it should have a prominent place, especially at the intermediate schools. In fact, the Panel operated under the assumption that students would arrive at the senior schools already well-grounded in history.[503]

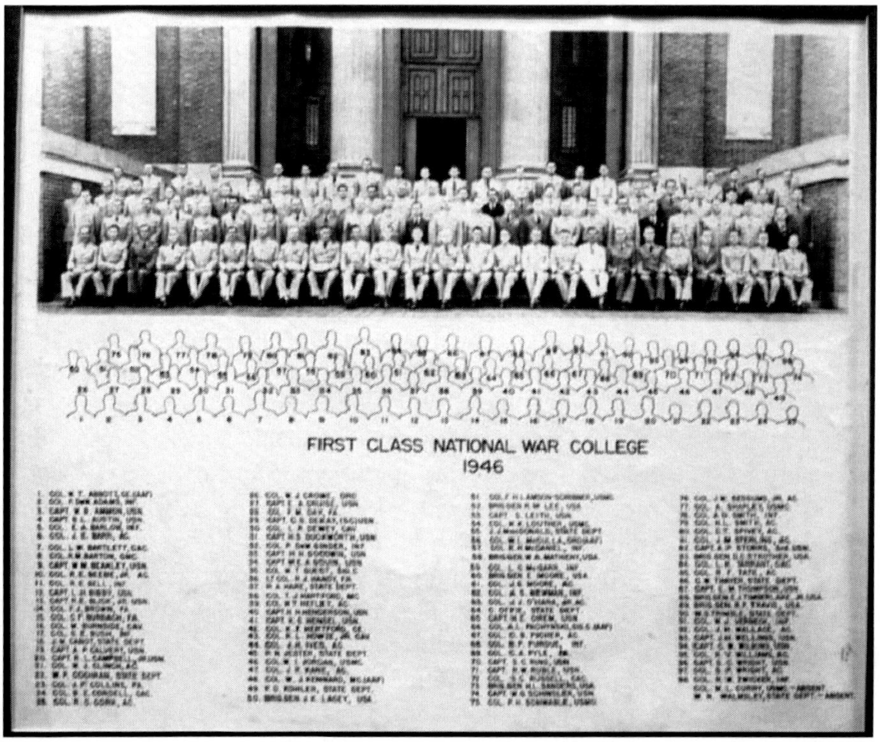

First Class National War College (1946).

Strangely, there is no real mention of history in the OPMEP. There are no Joint Learning Areas and Objectives for military history, even for joint operations. It is arguable that the inclusion of military history as a specific curricular area for evaluation by the PAJE would provide insulation from consideration for cutbacks to accommodate new material. Still, there are no established criteria for measuring student understanding or appreciation for what many deem to be the foundation of PME.

While history courses are primarily elective, the schools integrate military history throughout their core curricula and underscore special history-related activities such as staff rides to historical battlefields. School leaders consistently emphasized history's importance as the underpinning of their curricula.[504] The Marine C&SC goes as far as including the study of history in its mission statement with the preamble, "Informed by the study of history and culture. . . ."[505]

The Subcommittee heard complaints from faculty in more than one location that the classroom hours needed for adding emerging topics, such as irregular warfare or science and engineering, had come largely out of the previous treatment of history.[506] Given the dispersion throughout the coursework, it is difficult to gauge the total number of curriculum hours spent teaching military history. Nonetheless, there appears to be sensitivity among history faculty in some schools over a shrinking portion of the curriculum being spent on the study of history.

Observation: There is no military history requirement in the OPMEP. There is a perception among some faculty that the trade-offs between enduring and emerging subject matter are coming at the expense of providing an adequate foundation in history for the students.

OFFICER PROFESSIONAL MILITARY EDUCATION (PME) SCHOOLS: FLOW CHART

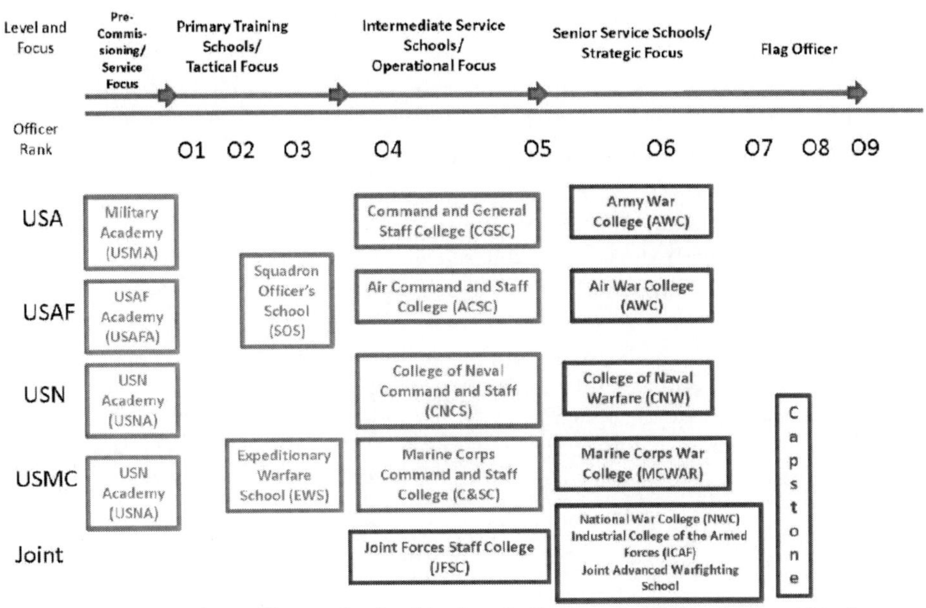

Note: There are many other military schools; this chart includes only PME schools within the scope of the 0&1 Study.

INSTITUTIONAL ISSUES

This section of the report begins with a review of professional military education (PME) organizational structures starting with the Office of the Secretary of Defense (OSD) and Joint Chiefs of Staff and moving to the joint institutions. The report then briefly reviews service PME structures and issues with service PME institutions. Next, the report highlights human resources with brief portions on diversity and quality in general before discussing senior leaders, faculty, staff, and students. Finally, this section of the report briefly reviews several issues related to material resources. The faculty portion of this report is significant. The Subcommittee notes, as did the Skelton Panel, that faculty is the bedrock of the PME enterprise. Faculty issues, therefore, warranted most of the Subcommittee's attention in this section of the report.

Organizational Structures

> *"What is the appropriate [PME] organizational framework to provide control and oversight?"*[507]
>
> **Jeffrey D. McCausland**

Department of Defense and Joint Chiefs of Staff Roles in PME

The Secretary of Defense (the Secretary) has largely delegated direction of professional military education to the Chairman of the Joint Chiefs of Staff (CJCS), who is said to "own" the joint aspects of PME at each institution.[508] During the course of this study, some experts questioned whether the Secretary of Defense's staff should take a more active role. Many other experts, staff, and faculty members whom the Subcommittee interviewed questioned whether the CJCS, given his demanding responsibilities, had the time to give more attention to the joint PME (JPME) system in general and the joint schools in particular. Still others questioned whether the Service Chiefs gave enough attention to their PME institutions.[509]

The Office of the Secretary of Defense

The Secretary is legally directed to "implement a comprehensive framework for the joint professional military education of officers." More specifically, the Secretary:

> ...with the advice and assistance of the Chairman of the Joint Chiefs of Staff, shall periodically review and revise the curriculum of each school of the National Defense University (and of any other joint professional military education school) to enhance education and training of officers in joint matters. The Secretary shall require such schools to maintain rigorous standards for military education of officers within the joint specialty....[510]

The Secretary of Defense shall require that each Department of Defense school concerned with professional military education periodically review and revise its curriculum for senior and intermediate grade officers in order to strengthen the focus on –

> (1) joint matters; and
> (2) preparing officers for joint duty assignments.[511]

The Secretary is also statutorily required to "implement a three-phase approach to joint professional military education" to include:

> ...a course of instruction, designated and certified by the Secretary of Defense with the advice and assistance of the Chairman of the Joint Chiefs of Staff as Phase I instruction, consisting of all the elements of a joint professional military education...in addition to the principal curriculum taught to all officers at an intermediate level service school.[512]

JPME II instruction is required to be “taught in-residence” at Joint Forces Staff College (JFSC) or senior-level service schools, and JPME III must be offered via the Capstone course.[513] Officers must complete JPME I before proceeding to JPME II.[514] The law also specifies that the Secretary will define the curriculum for JPME II to focus on joint operational expertise and perspectives and joint warfighting skills “to adequately prepare students to perform effectively in an assignment to a joint, multiservice organization” and “so that students progress from a basic knowledge of joint matters learned in JPME I instruction to the level of expertise necessary for successful performance in the joint arena.”[515]

The law also requires the Secretary to submit an annual PME report to Congress with information on the Department as a whole, each of the services, and each reserve component. The report must include the number of officers who completed a JPME II course, but who were not selected for promotion, and the number of officer students and faculty members that each service assigns either to the schools of the other services or to the joint schools.[516]

Finally, the Secretary funds the activities of the JPME schools, while funding for their infrastructure is programmed by the service that hosts them (the Army at Ft. McNair for the Industrial College of the Armed Forces (ICAF), National War College (National), and Capstone; the Navy at Norfolk for JFSC).[517]

The Under Secretary of Defense for Personnel and Readiness (USD(P&R)) has played a limited role in the direction of PME and JPME; however, it is the one office within the OSD that has the largest role. The small USD(P&R) staff under the Director of Officer and Enlisted Personnel attends the Military Education Coordination Council (MECC) as observers and supports the Process for Accreditation of Joint Education (PAJE) visits to the intermediate- and senior-level schools. Otherwise, USD(P&R) provides oversight of three specific programs that educate military professionals: officer graduate education, training with industry, and fellowships.[518] OSD oversees the Joint Staff work on the Joint Qualification System and policy. USD(P&R) works with the Joint Staff directorates for Personnel (DJ-1) and Training and Education (DJ-7) to ensure compliance with the law and OSD policy.[519]

Although OSD’s practical role has been relatively unpronounced, the April 2008 Secretary of Defense *Guidance for the Development of the Force* (GDF) calls on USD(P&R) to: lead an effort to revise JPME content to address the full range of domestic and overseas operations; to include civilian education; and, to prepare officers and DOD civilian personnel for joint, combined, and interagency operations.[520] The GDF has replaced the *Strategic Planning Guidance*, *Transformation Planning Guidance*, the *Posture Guidance*, the *Science and Technology Posture Guidance*, and several other DOD guidance documents.

Historically, OSD had the option to play a larger role for in PME. In the 1960s the Deputy Assistant Secretary of Defense (DASD) for Education, in addition to other responsibilities, had specific responsibility for PME.[521] A Ph.D. Principal Assistant for Education Programs and Management Training was subordinate to this DASD. In the late 1990s, the Secretary established the office of the Chancellor of Civilian Education and Professional Development, which was made up of three senior civilian educators (Ph.D.s) and a senior civilian with a doctorate (Senior Executive Service 3). This office did not have statutory or policy-based responsibility for PME. However, the chancellor’s office was located at National Defense University (NDU), and the chancellor attended MECC meetings and participated in PAJE visits on his own initiative.[522] The office was eliminated in 2004.[523] Although some experts believe that an OSD chancellor’s office should be re-established to direct PME efforts, all five senior PME school leaders disagreed with that proposal, to some extent.[524]

Finding: **The Guidance for the Development of the Force gives the Under Secretary of Defense for Personnel and Readiness (USD(P&R)) a role in PME. Specifically P&R, with the Joint Chiefs of Staff, is to revise JPME content to reflect new guidance in the Capstone Concept for Joint Operations (CCJO) and Joint Operating Environment (JOE). While the Subcommittee does not attribute any specific shortcomings to current efforts, the PME system could benefit from stronger civilian leadership, particularly in developing plans, policies, and programs.**

Recommendation: **USD(P&R) and the Joint Staff should brief the House Armed Services Committee on the curricular revision of joint content as called for in the Guidance for the Development of the Force.**

The Chairman of the Joint Chiefs of Staff

By law, the CJCS formulates policies for coordinating military education and training.[525] The Director of the Joint Staff is the CJCS' designee to manage the higher-level responsibilities for JPME by using the MECC, while the staff under the JCS/DJ-7, who in his role as the Deputy Director of the Joint Staff for Military Education (DDJS-ME) has responsibility for training and education, manages the daily business of military education policy.[526] The CJCS' *Officer Professional Military Education Policy* (OPMEP) also describes the roles of DJ-1 Director for Manpower and Personnel, DJ-5 Director for Strategic Plans and Policy, U.S. Joint Forces Command J-7, and the president of NDU.[527] The DJ-7 training and education staff is also responsible for doctrine and other programs.

Although the Education Division of the DJ-7 is charged with managing PME academic policy, it does not manage personnel billets designated for educators or academics.[528] In addition, the one-star DDJS-ME and his staff do not direct the JPME system or institutions as they are only in a position to "suggest" and "recommend."[529] On the other hand, the DDJS-ME can recommend to the Director of the Joint Staff and the CJCS that JPME accreditation should be withheld if an institution does not fulfill its OPMEP responsibilities. In addition, the Joint Chiefs of Staff Education Division staff largely comes from a doctrinal background, and, while this background is important, most doctrine focuses on what works now, rather than looking ahead.[530] While PME covers both doctrine and anticipates the future, PME should be weighted to provide the education officers require for confronting uncertainties in the future.

Finally, the DJ-7 does not appear to be in a position to take a comprehensive view of this very complex system that links training, education, and assignments both to the individual schools and to positions officers are sent to after they graduate from these schools. The Subcommittee has attempted to do this and realizes just how large a task it is. The JCS/DJ-7 staff implements the OPMEP, manages the MECC, and solicits and mediates decisions on special areas of emphasis.[531]

Finding: **USD(P&R) and JCS/DJ-7 do not have educators or academics involved in directing the JPME enterprise.**

***Recommendation*: The staff that administers JPME and PME for the Secretary and CJCS should be balanced between those who have academic credentials or educational experience and those who have operational backgrounds.**

The processes that the DJ-7 is responsible for largely occur in a closed system conducted primarily by the leaders of the schools. Faculty members from several schools inspect a school, and in turn, their institutions are inspected by faculty from their counterpart institutions. In other words, the "operators," or those who employ PME graduates, the Combatant Commands (COCOMs), are not directly represented. Possible issues that arise were highlighted in the previously mentioned Fenty Study. The Subcommittee would contrast the Military Education Coordination Council's closed system with the Air Force's Learning Committee, which includes the vice commanders of all its major commands, as an example of a relatively open system.[532] In this case, the employers of the graduates can bring their requirements to the educational forum where additions to the curriculum are debated and balanced. For instance, a vice commander may ask that the strategic implications of cyber war be studied in relation to "just war" theory. This goes beyond the annual call for special areas of emphasis that the Joint Staff puts out to the services and COCOMs.

***Finding*: The Military Education Coordinating Council could be more effective as a more open system.**

***Recommendation*: The Department and CJCS should strengthen the MECC and MECC working group with the formal inclusion of COCOM representatives at the level of the other participants.**

The Subcommittee did not examine alternative possibilities for senior-level management of PME in detail and has every confidence that the CJCS has great concern for the educational enterprise. However, while this issue arose during interviews, briefings, and hearings, there was no consensus on where the various top-level and daily management responsibilities should be placed. During the Subcommittee's first PME hearing, witnesses argued that the system needs better top- level management. Dr. Alexander Cochran proposed that a Chancellor of Military Education Office be established within the OSD to provide a higher level of oversight of and support for PME.[533] Dr. Janet Breslin-Smith called for the CJCS to "reclaim ownership," particularly of the joint schools. She views the joint schools as "orphaned" by senior leadership that is "too detached," particularly in the budget season but also as to content and stature.[534] Other experts and practitioners have asked why Joint Forces Command does not direct the JPME courses or control NDU or, at least, Joint Forces Staff College.[535] Still others take the position that the president of NDU is better equipped to direct the joint PME schools and courses for the CJCS, and possibly joint education, more generally since the focus of that position is exclusively on education.[536] In other words, some experts are looking for a three- or four-star officer (other than the Director of the Joint Staff) or a civilian- equivalent JPME leader, who can manage the system comprehensively and is focused solely on the operations of the joint educational enterprise on a daily basis.[537]

The Skelton Panel went even further to include all PME. The 1989 report observed that the DJ-7:

> …who currently has this policy responsibility, also has other responsibilities—war plans, interoperability, and joint doctrine—that are so large that he has limited time to focus on important educational issues. In fact, the senior Joint Staff position with full-time education responsibility is at the colonel/Navy captain branch chief level.[538]

In the intervening 20 years, little has changed in this regard. It may be time to reconsider the Skelton Panel's recommendation for a full-time Director of Military Education apart from being an additional set of responsibilities for the Director of the Joint Staff and the DJ-7 as currently assigned. The Panel called for this senior officer to have responsibility beyond developing, accrediting, and monitoring joint education in service and joint PME schools. The Report called for this officer to have strong academic credentials and to be charged with establishing a coherent framework for all intermediate and senior PME schools as well as coordinating military education overall.[539] Such a change could also engender reconsideration of the organizational location for the daily management of JPME where the constant operational focus could be on academic programs.

A comparison can be made to the four-star Army Training and Doctrine Command (TRADOC) commander and the four-star Air Force commander of Air Education and Training Command (AETC) who, despite differences in their portfolios, are both charged with directing their services' education systems. The current organizational system, which calls on the CJCS and the Director of the Joint Staff (who have other significant daily concerns) to manage JPME, has led to a mechanistic implementation of the OPMEP and approach to special areas of emphasis. The organizational system has also led to a piecemeal approach to requests for changes to the larger system. Sometimes these requests for changes exclude consideration of the other linkages such as those to the Joint Officer Management system referred to earlier. Only very rarely, if ever, can the Joint Staff try to conduct a comprehensive review and analysis of PME.

***Finding*: Twenty years later, there is no dedicated full-time director of military education that could respond in a comprehensive way to the spirit of the Skelton Panel's recommendation.**

Joint PME Institutions

All of the PME institutions now present some form and level of joint education. The Subcommittee notes, as some witnesses suggested, that the leaders of the PME institutions at different levels have no regularly scheduled, formal fora for coordination and communication.[540]

The leaders of the senior PME schools do see each other at various fora informally several times a year, according to the commandant of the Army War College.[541] The Military Education Coordination Council has a specific purpose that does not lend itself to these kinds of interactions.

***Finding*: The PME enterprise does not have regularly-scheduled, formal fora for coordination and communication outside specific OPMEP discussions in the Military Education Coordination Council.**

***Recommendation*: A regular, formal forum for PME school leaders should be established to share common concerns and best practices.**

President Harry S. Truman presented diplomas to the National War College Class of 1949.

In addition to how the JPME enterprise is organized, each JPME and PME institution is internally organized very differently. The joint schools report to the CJCS through a three-star NDU president. In addition, NDU established a new civilian provost/vice president for academic affairs position. The recently departed provost had an extensive academic administrative background but no military background. The person next hired will hold only the title of vice president for academic affairs. NDU, founded in 1976 to centralize oversight of joint education institutions, has since become home to what one informed observer referred to as a "dizzying proliferation" of schools, institutes, and centers.[542] What started as a three-college university now has command of 14 organizational entities as well as administrative control over a number of regional centers.[543]

This study focused on the three institutions within NDU that grant JPME credit; however, it is important to ask whether the focus and the "span of control" of the university are appropriate. Some experts and current faculty and staff at the three JPME institutions under NDU have, over time, begun to feel that their schools are "orphans" within the larger university system.[544] According to some professors and staff, the three original colleges believe that they do not receive adequate attention and support from the university because the NDU staff is preoccupied with many other entities and activities.[545]

The three colleges and the Capstone course under NDU are each organized differently. ICAF has a two-star commandant, a retired colonel as dean of students (until very recently this position was filled by an active-duty colonel), and a civilian Ph.D. dean of faculty. National has a two-star commandant, a military colonel Ph.D. as dean of faculty, and a retired colonel as chief of staff and dean of students. The latter position was placed above the dean of faculty in the chain of command in the role of chief of staff.[546] The JFSC has a two-star billet for the commandant, a senior civilian Ph.D. as dean of academic affairs, and 0-6 directors of Joint and Combined Warfare School (JCWS) and Joint Advanced Warfighting School. These two JFSC schools do not have their own deans.[547] Capstone's new director is a retired rear admiral who has a small administrative staff rather than full-time faculty. A standing group of senior mentors (retired three- and four-star flag officers) helps the staff with each session of this flag-level course. The Capstone director reports directly to the president of NDU.[548] Some of the senior mentors have been with the program for a long period of time.[549]

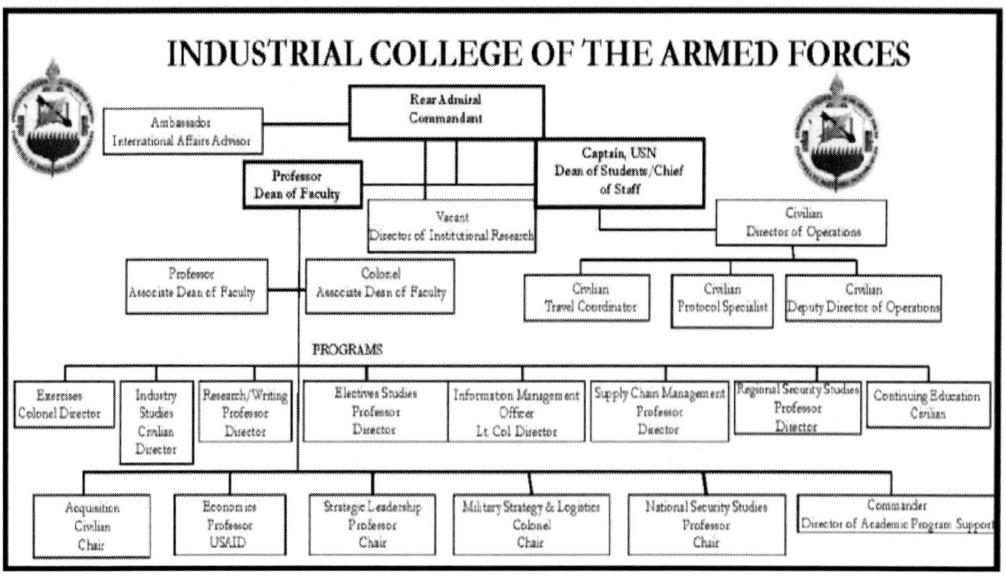

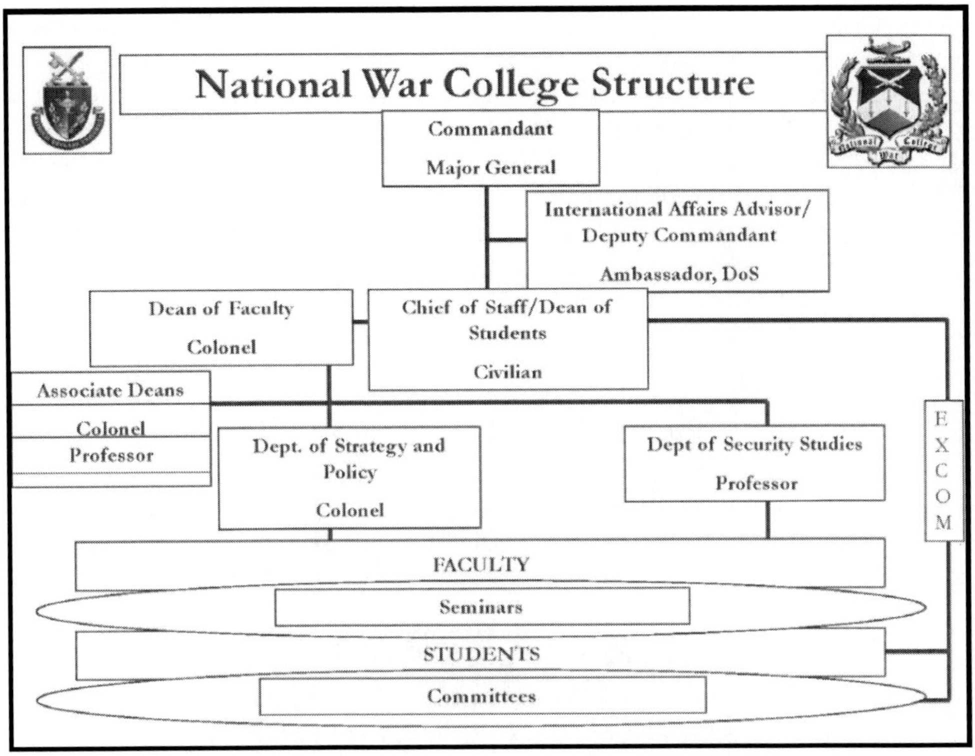

When ICAF's and National's dean of academics positions have been filled by military officers, there has been rapid turnover. Sometimes the incumbent does not have the best academic credentials for service at NDU. For example, some have had a Ph.D. in the sciences rather than in political science, history, or national security studies. According to Subcommittee witnesses and interviews, more continuity and the right credentials are the key to success for these institutions.[550] A civilian academic of appropriate stature, in the right field, instills confidence in the faculty as their dean and advocate to the administration. A military officer on rotational assignment, as dean of students, enhances the balance between military and civilian perspectives among the senior leaders. Interjecting a non-academic in the academic chain of command between the senior academic leader, the dean, and the commandant is potentially problematic.

***Finding*: The dean of academics and dean of student positions at ICAF and National are not optimally situated within those organizations. The dean of students, dual-hatted as the chief of staff, should not have academic responsibilities in the chain between the commandant and the faculty.**

***Recommendation*: For continuity purposes and experience reasons, it is the right time to review the dean of students and academic dean positions at ICAF and National to set the appropriate general credentials and tour length.**

Service PME Organization

The service schools, the Service Secretaries, and the Service Chiefs have a variety of relationships with the CJCS relative to PME, and they have various degrees of personal involvement in their senior service schools.[551] The reporting chain for the presidents and commandants of the PME institutions is significant in that it can show how highly education and educational leadership assignments are valued in the organization. It also can help ensure that education retains a distinct and complementary value in relation to training. Finally, it is important because of resourcing issues. The more senior the attention, the better chance there is that any function will be well-resourced.

The Army Chief of Staff has appointed the TRADOC commanding general to direct the Human Resources Enterprise. This four-star general coordinates with the Department of the Army and Army Staff G1 (Personnel) and G3/5/7 (Operations, Plans, and Strategy) at Headquarters, Department of the Army. In addition, he coordinates with the Human Resources Command, the U.S. Army Reserve Command, and the commanding generals of the Army National Guard and the National Guard Bureau among others. For instance, the commanding general of TRADOC has led the development of the officer education continuum including the connection between various phases of training and education and their alignment with the officer personnel system. The TRADOC commanding general has direct command over the Combined Arms Center, including the Command and General Staff College (CGSC), as well as the Army War College.[552]

The United States Air Force Thunderbirds help the cadets celebrate their graduation with a flyover and a post graduation airshow during the 2009 United States Air Force Academy graduation ceremony.

A four-star general commands the Air Force's Air Education and Training Command (AETC). While he has command over the Air University and, therefore, Air War College, Air Command and Staff College, and Squadron Officers College, he does not have the charter to bring together PME and the personnel system which is directed by the respective commanding generals of the Air Force Personnel Center (AFPC) and the Air Reserve Personnel Center (ARPC, which directs both Reserve and Air National Guard personnel affairs). AETC, AFPC, and ARPC all coordinate their policies and activities related to personnel, education, and training with the Headquarters, Air Force's Deputy Chief of Staff for Personnel (A-1).[553]

The Navy's Naval Education and Training Command is commanded by a four-star admiral separate from the Bureau of Naval Personnel. Neither has a direct role in PME. The president of Naval War College (NWC) reports directly to the Chief of Naval Operations.[554] The Marine Corps manages its personnel through Manpower and Reserve Affairs. The Marine Corps appoints an active duty two-star general or a retired senior officer as president of Marine Corps University (MCU). He reports through the two-star commander of the Training and Education Command and the three-star commander of Marine Corps Combat Development Center to the Commandant of the Marine Corps.[555]

Observation: Although the Subcommittee found a disconnect between the personnel (selection and assignment) system and PME, the Army is trying to remedy this disconnect by making TRADOC the executive agent for a new Human Capital Enterprise that includes both. That said, the Subcommittee finds that it is too soon to evaluate the effectiveness of this approach.

Dr. James Carafano testified to the Subcommittee that he thought it a "travesty" for the Army that the Army War College commandant has been subordinated under the commander of TRADOC.[556] Dr. Richard Kohn, in agreement, indicated in his testimony that TRADOC does not have "education" in its name.[557] Carafano said, "[Congress has] to legislate that the Service school belongs to the Chief or the Commandant....It is your college. You have to keep it."[558] The existing arrangement does not allow the Army War College commandant or commander of the Combined Arms Center to directly petition the Chief of Staff of the Army with concerns. For some, this does not connote the level of value they think the position, the school, and PME warrant.

General Martin Dempsey and General George Casey, on the other hand, have sought a consolidated and a comprehensive approach to Army Leader Development that has its own advantages.[559] For instance, the TRADOC commander can take an integrated approach to various phases of education and experience. Others agree with Carafano that all the joint PME school commandants should report to the CJCS, in the case of the joint schools, or to their respective Service Chief if they command a service PME school.[560]

Jeffrey McCausland, in his study of the war colleges in 2005, said the issue of directly reporting to the Service Chief is "of critical importance" and education should be separated from the training function.

> This importance derives from the rapid rotation of commandants at all the schools—which tends to erode the spirit of educational autonomy over time—and whether the college focus on education can be successfully accommodated in a structure that is deeply and zealously imbued with the philosophy and culture of training. ...[A]nd giving oversight to service training headquarters also has the potential to impart excessive service biases in curriculums.[561]

Resources will be addressed later, but McCausland believes that the "locus of control and oversight also has an immediate impact on budget allocations and external taskings."[562]

Service Individual Institution Organizational Issues

At the Service individual institution level, some organizational arrangements may be less than optimal for the PME mission or for best use of resources. The Subcommittee does not find any shortcomings directly linked to organization, and will not dictate any "ideal" structure, but some arrangements may warrant reconsideration by the institutions themselves. Air Force schools stand out among service schools.

The Air Force PME institutions have a complicated structure, according to faculty members and leaders of those schools. As explained above, the three-star Air University commander reports to the four-star commander of Air Education and Training Command and has a broad portfolio of professional and graduate education institutions under his purview. Included are the three colleges under the Spaatz Center for Officer Professional Education in Alabama. The Spaatz Center is commanded by a major general assisted by a civilian chief academic officer. The Spaatz Center commander and chief academic officer are also dual-hatted as the leaders of Air War College (AWC), but the AWC commandant also has an academic dean for the in-residence AWC course. Air Command and Staff College has a one-star commander (dual-hatted as the vice commander of the Spaatz Center) with both a chief academic officer and dean. Finally, there is a colonel who commands Squadron Officer College (SOC) who also has a chief academic officer and dean. SOC includes the Air and Space Basic Course (ASBC) (for second lieutenants) and Squadron Officers School (for captains).[563]

Several Air University organizational issues came up in the course of the Subcommittee's study. First, Dr. Richard Kohn recommended in his testimony to the Subcommittee that the Air Force consider merging the faculties of the staff college, the war college, and the School of Advanced Air and Space School so that subject-matter expertise could be shared and leveraged to strengthen instruction in all three schools, using the example of civilian universities and the Naval War College.[564] The Subcommittee also observed that the wargaming and research arms of the university could be better integrated with the colleges to enhance educational synergy for the students there.[565]

Second, the Joint Staff's report on the 2008 accreditation of the AWC cautioned that although the AWC's joining of Air University's Spaatz Center brings some benefit, AWC "should maintain its uniqueness and identity as a Senior Level College....The AWC should make a deliberate effort to keep members of the Education Support Squadron (ESS) of the Spaatz Center thoroughly indoctrinated into the individual mission of the AWC."[566] It was also noted that AWC has had hiring delays by having to go through two levels of authority for these decisions, which is addressed in the faculty section of this report.[567] (The Spaatz Center organizational chart is below and charts for the three component schools discussed here are at Appendix C.)

Finally, the McCausland Study, reinforced by the Subcommittee's observations, pointed out the problem with having both deans and CAOs: "each college should also have a single senior officer or civilian educator serving as both Dean and Provost to underscore the point that education, not training for the issue *du jour*, is the fundamental imperative of its institutional mission."[568] Without a specific rationale and delineation of roles and responsibilities for each, the faculty, staff, and students, as well as outsiders, are confused by this organizational structure, not knowing who is empowered to make what decisions.[569] In addition, it may not be resource-effective. If the problem is the continuity of military deans, then perhaps a better solution is a controlled assignment of three or more years, or the hiring of a civilian academic as dean.

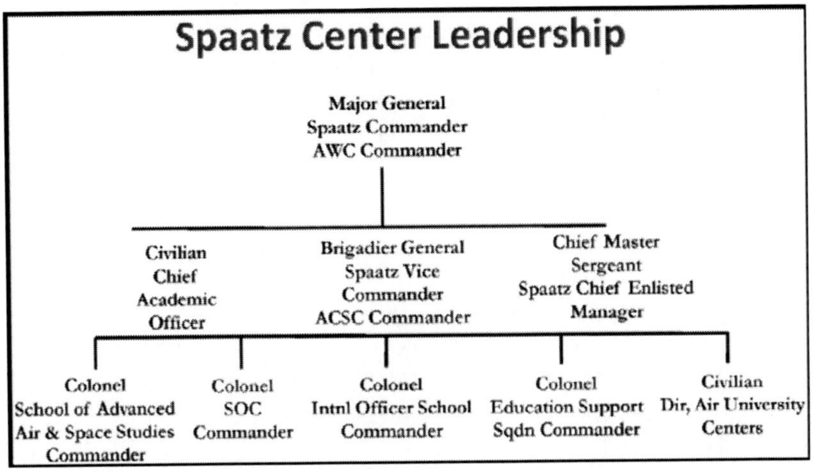

***Finding*: While some of the larger schools or universities have an academic vice president or provost, Air University (AU) has six academic leaders for three schools. The Chief Academic Officer (CAO) of AU, who seems to act as university provost, is also dual-hatted as the CAO of Air War College and serves there with a civilian academic dean. Air Command and Staff College and Squadron Officer College also each have a CAO and dean. The CAOs tend to be civilian academics while the deans are either military or civilian academics. Without a clear delineation of responsibilities, the organizational structure seems confusing and resource intensive.**

***Recommendation*: Air University should clearly delineate the responsibilities of both CAOs and deans and consider whether a civilian dean with credentials in a discipline related to national security might be most appropriate as the academic leader of the in-resident PME courses.**

Human Resources

> *"Fundamental to the development of the U.S. officer corps is quality professional military education (PME). ... Because education is an investment in our nation's future, the services must be willing to sacrifice some near-term readiness for the long-term intellectual development of their officers. Only by accepting these sacrifices will our officers have the intellectual talents to respond to the demands of their profession, especially in major crises and wars."*[570]
>
> **The Skelton Report, 1989**

People are critical to the PME enterprise. Together with pedagogy, the Skelton Panel considered the senior staff, faculty, and student bodies of the various PME institutions to constitute the "bedrock" of quality for PME.[571] At the same time, perhaps the most difficult challenge to the PME institutions is measuring the quality of the people involved in the PME enterprise.

The Subcommittee found that the persons involved in the PME enterprise generally meet high qualitative standards. However, challenges to standards should be addressed either to avert or correct qualitative decline. This section of the report will discuss diversity as an essential element of quality as well as examine the categories of people involved in the PME system including senior leaders, military and civilian faculty, and support personnel. Students are addressed in a succeeding section. Before addressing these groups that play important roles at the PME institutions and the issues they face, the report discusses the other two issues: the role of diversity and quality in general.

Diversity

Military leaders at all levels must understand their diverse work force, the balance of civil- military relations in the United States, the partners and allies with which their forces operate, and the adversaries they face. The leaders at each PME school, as well as the CJCS and the Service Chiefs, emphasize how important diversity is to education. Therefore, the specific mission of the PME institutions demands that leaders pay attention to the diversity of senior staff, faculty, and students.

Because diversity matters in terms of the importance of a variety of perspectives on national security, schools must search for more diversity among qualified senior leaders and faculty for PME institutions. Human resource professionals have learned that diversity does not automatically follow from traditional means of advertising or recruiting for openings.[572] These traditional recruiting methods do not always reach or attract diverse candidates. Non-traditional, more creative methods at more locations are required.[573]

The Subcommittee is encouraged by the commitment to diversity voiced at every level of education and among the ranks of senior leaders. However, PME institutions' efforts to recruit both diverse military faculty and civilian professors often lag behind their aspirations.[574] At the primary-PME level most faculty are military and therefore reflect service demographics. At the intermediate and senior levels, while the institutions all want more diversity, most of the senior staffs and faculties still include few women and fewer people of color or varied ethnic backgrounds.

A diverse military force must be grown from the bottom up. The three service academies seem to have the greatest expertise with student diversity issues, as they recruit students from demographic pools outside the military. They acknowledge doing best both with diversity and quality when their outreach efforts to minority communities, including hosting summer camps, are fully funded.[575] Military students at intermediate- and senior-level schools will reflect the composition of personnel recruited 10 and 20 years earlier.

Civilian faculty can be recruited from more varied sources. For instance, in addition to advertising in *The Chronicle of Higher Education*, schools might also reach out to historically black colleges and universities, tribal colleges, or the American Association of University Women. The Naval War College has issued a new civilian faculty recruiting policy focusing on non-traditional candidate sources that will help in this regard.[576]

***Finding*: The PME institutions agree on the importance of promoting diversity among qualified personnel. However, leaders of these institutions admitted they did not always undertake the recruiting efforts that might yield greater diversity. The Subcommittee understands that the faculty and senior staffs are partially a reflection of both the recruiting population of current and retired senior military officers and the**

population of professional educators who focus on military and national security matters.

***Recommendation*: The Subcommittee encourages PME institutions to take further imaginative and creative steps to contact and attract qualified, diverse senior leaders and faculty members such as recruiting at non-traditional and diverse institutions and advertising in periodicals aimed at women and minorities. Other PME institutions could use the new Naval War College civilian faculty recruiting policy as a benchmark.**

Quality in General

The Subcommittee considered the central question of how PME institutions attract, develop, and retain the highest quality senior staff and faculty. The Subcommittee found that some issues were associated with civilians, others with military personnel, and still others pertained to both. Many faculty issues are interrelated and complex. Senior staff issues at the primary, intermediate, and senior schools, both joint and service, focus on qualifications and the lengths of leaders' PME assignments as matters affecting the stability of institutions. Civilian staff and professors were most concerned about tenure. To military faculty, balancing stability against promotability was very important. For civilian professors at some institutions, there were similar concerns about promotability within institutions and marketability when they leave government employ.[577] Issues that concerned both military and civilian faculty include balancing requirements for teaching and curriculum development; research and publishing; and service and outreach. The specific balance of these three areas sets some schools apart. In general, PME institutions, if they are pressed for resources, focus most on teaching and service. However, this balance can also be altered by what is expected of different kinds of faculty (military and civilian, junior and senior) in an academic and military division of labor. Two other issues common to military and civilian faculty members related to publishing and intellectual property (copyright) and standards of academic freedom.[578] Common issues are addressed at the end of the faculty section of this report.

Senior Staffs

Presidents, Commanders, Commandants, and Directors

Quality leadership at PME institutions starts at the highest levels. The Skelton Panel reported that commandants and presidents were second only to faculty in importance for maintaining quality PME and that they "should play a significant role in guiding the curricula and mentoring the faculty and student body."[579] The Subcommittee found that, while there is no question that the current senior leaders of the PME institutions are quality officers, there is continuing debate about how school presidents, commanders, commandants, and directors should be selected and how long they should remain in their positions.

The Skelton Panel maintained that the flag officers selected for these positions should have "operational credibility, academic credentials, a superb intellect, and must be seen by the student body as having the highest standard of integrity….[and] the billet must be viewed by the service chiefs as an assignment of major importance."[580] The Skelton Panel also asserted that only the CJCS for a joint school, or a Service Chief for a service school, should make the selection.[581] The Panel suggested five selection criteria: (1) a strong academic inclination; (2) upward career potential; (3) willingness to devote a minimum of three years to the institution; (4) current operational knowledge; and, (5) the ability to establish sound rapport with the student body. The Panel also suggested that these senior leaders should have some type of educational (as distinct from training) background and should teach "at least one or two courses."[582] The witnesses at the Subcommittee's first PME hearing agreed with the Skelton Report, specifically noting that running a hybrid academic and military institution has very unique challenges.[583] Dr. Janet Breslin-Smith, Dr. Alexander Cochran, and Dr. Richard Kohn were unanimous in supporting the position that the commandants should have a longer tenure and that they should have an educational background in addition to operational experience. Breslin-Smith argued that the CJCS should review the criteria for senior billets and that the person selected needs to be committed to, and intellectually involved in, teaching and engaging the students in strategic dilemmas. Kohn suggested that if a suitable military leader could not be found, a retired officer or a civilian would be better than a less qualified, active-duty officer.[584] Other current and former faculty members also questioned the qualifications of senior military PME leaders, particularly if he or she remained aloof from daily educational activities but exercised a heavy hand in controlling the academic aspects of the school.[585] Dr. Williamson Murray suggested that the selection of leaders without academic expertise indicates that service leaders do not regard PME as sufficiently important.[586] However, persons holding these positions have sometimes earned promotion to the next higher rank.[587] Accordingly, these PME leadership assignments are not necessarily perceived to be "dead end" or retirement positions.

Leaders of the senior PME institutions argued that academic expertise is less important than recent operational experience and leadership skills.[588] As a result the services generally treat these assignments as two-year O-6 or flag-officer command positions. The majority of these leaders do not see an overriding requirement for experience in or talent for education for what they see as a military command.[589] While the ACSC commandant maintained that "there is no academic qualification for the commandant position," the other intermediate PME school leaders voiced the opinion that a balance of academic and operational background should be weighed in the selection. The CGSC deputy commandant opined that his position, "requires less focus on academic experience and more on operational leadership skills and experience," and added that recent operational experience was necessary for his role in reviewing the contents of the curricula.[590]

> The Joint Staff's position is clear,
>
> Regarding Senior Schoolhouse Leadership, each selection is carefully weighed. Service chiefs bear responsibility for the choices made at their schools. The Chairman selects the General/Flag Officer leadership for the NDU Schools and makes a recommendation to the Secretary of Defense regarding the officer to be the NDU President. NDU assignments rely on quality nominations from the Services.[591]

However, neither the Secretary of Defense nor the CJCS, nor the service secretaries or their chiefs promulgate specific criteria for selection to senior PME leadership.[592]

Finding: **No qualifications have been promulgated for senior PME leadership positions. Some of the services do look for academic or educational backgrounds; others consider these positions to be simply the equivalent of other flag-officer command or leadership positions. Although operational experience is important, even in times of war**

a "strong academic inclination" is still an important qualification for PME senior leadership.

***Recommendation*: The CJCS and Service Chiefs should establish general criteria or credential requirements, such as demonstrated educational and academic interest, for selecting PME directors, commandants, commanders, and presidents to be applied at the Chairman's and Service Chiefs' discretion.**

Senior staff tour lengths are also an issue of debate. Appropriate tenure is critical for continuity. As discussed above, if the services view these senior leadership positions as simply another assignment for flag officers, their tenure will generally be two years or less. The Skelton Panel recommended senior leaders serve at least three years in their posts, except during times of great transition when four to five years would be preferable.[593] Witnesses at the Subcommittee's first PME hearing, interviews with current and former faculty members and commandants, and Dr. Jeffrey McCausland's 2005 PME study warned that short and unstable tour lengths for senior leaders have an adverse effect on PME institutions.[594] Although he favors three- to five-year assignments, Dr. Cochran goes so far as to suggest 10-year tour lengths. If commandants serve for longer than three to five years, he proposed they be given "tombstone" promotions to the next higher rank upon retirement as is done at the U.S. Military Academy (USMA) and U.S. Air Force Academy (USAFA) for department heads.[595]

Observation: Senior staff tour lengths and longevity issues may be in tension with promotability. Some schools or services choose to compromise between upward mobility for the right candidate, while some candidates choose not to maintain their promotability to stay in their positions longer. The Subcommittee finds no reason to disagree with the Skelton Report recommendation that tour lengths of three to five years are "about right" for PME leadership positions.

Provosts, Deans, and Senior or Chief Academic Officers

While the Skelton Report was silent on academic leadership, some variations on the foregoing debate also occur over the qualifications for civilian provosts, chief academic officers, and deans. Each of the services and schools takes a different approach to hiring academic leaders. Qualification issues include academic management experience, teaching expertise, credibility in one's field as it relates to PME subject matter, and experience in (or at least exposure to) the military. Some schools prefer to hire military officers with Ph.D.s but not necessarily in a discipline related to PME. For instance, deans might have a degree in physics rather than a national security-related field such as international relations. Other schools prefer to hire civilians with academic management experience. Some experts suggest that the best way to maintain high-quality, rigorous academic programs and to attract outstanding civilian faculty is to hire as dean an academic who is world renowned in the areas of strategy and national security. Civilian academics offer longevity and continuity, and they can have an exceptional understanding of faculty issues. They bring a different perspective from that of the military leader at the top of the institution. If civilians are hired into leadership positions at PME schools, however, they must also have an appreciation for the combined military and academic environment.

Outside experts and the leaders of the schools weighed in on the issue of the senior academic officers' credentials. Most agreed that the academic dean should have excellent academic credentials and teaching experience. Dr. Breslin-Smith testified, "the Dean of Faculty should have a Ph.D., prior teaching experience, and a commitment to the mission of the College."[596] Dr. Cochran suggested that deans, definitely civilians, should be chosen for their academic background in a PME- relevant field and stay in the position at least twice as long as the military head.[597] Dr. Kohn disagreed, saying the deans should be military officers with proper academic credentials chosen from the faculty after "demonstrated accomplishment as teachers, leaders, and scholars in their discipline or field."[598] The senior leaders at the schools take a variety of approaches. The ICAF commandant believes academic deans "should have extensive teaching and supervisory experience together with policy experience in the military or an executive branch agency."[599] The JFSC commandant wrote, "the Academic Dean should have a doctorate in a relevant field such as education or national security studies, have attained the rank of full professor, have had exposure to the military environment, have experience in higher education and administration, and have an extensive background in managing and supervision of large dynamic organizations." Marine Command and Staff College (Marine C&SC) recently named a civilian dean to preclude rapid turnover that the school experienced with military deans.[600]

***Finding:* The several PME institutions take a variety of approaches to hiring their most senior academic leaders.**

***Recommendation*: Whether military or civilian, deans should possess a Ph.D. or terminal degree (typically a law degree is considered "terminal") in a subject area relevant to national security and be well respected by their peers; should be at least familiar with the military, even if they have not served themselves; and should anticipate at least a three- to five-year tenure.**

Faculty

The Skelton Panel considered faculty "the determinant factor in quality education."[601] The Panel reported, "The importance of a competent, credible, and dedicated faculty to both the fabric and the reputations of our PME institutions cannot be overstated."[602] Today, all of the PME schools report having difficulty recruiting a sufficient number of the highest quality military faculty.[603] Some schools also report challenges recruiting civilian faculty.[604] Recruiting issues include increasing diversity, type and level of qualifying degrees held, opportunities for command or leadership positions, and competition with civilian schools.

Experts have suggested that there is a need for even greater diversity among the faculty, including professors from different backgrounds and an increase in faculty with social science degrees.[605] For military officers, the schools have to compete for the best candidates with command, key field or operational assignments, and joint staff assignments. For civilian academics, the schools have to compete with the nation's most prestigious universities to build a highly respected faculty.

The Skelton Report divided military faculty into three categories.[606] First, military operators are those who have recent and relevant operational experience and are leaders who may or may not have advanced academic credentials or a talent for teaching. Second, there are military specialists such as foreign area experts or strategists who may have advanced degrees but no teaching experience. Their area of expertise directly relates to their teaching assignment. And, third, military academics or educators are those who have risen through the military ranks but who also have advanced degrees, subject-matter expertise, and teaching experience.[607] The Panel believed that military faculty should have at least a master's or professional degree and that at the senior school level a doctorate was desirable.[608]

Civilian instructors (hired under the authorities provided to the Secretary of Defense under Title 5 or Title 10, U.S. Code) also have various backgrounds. When referring to civilian employees or faculty, Title 5 refers to employees hired under the General Schedule or GS system. Title 10 refers to those civilians hired under special authority given to the Secretary of Defense and the Service Secretaries to hire "professors, instructors, and lecturers" in the numbers they consider necessary for the JPME schools, the service officer in-residence PME courses of at least 10-month duration, and the three service academies.[609]

The Skelton Report stated that all civilian instructors, at least at senior schools, should have doctorates and should include a mix of types.[610] First, some civilian faculty members are retired military officers who have advanced degrees as well as leadership and operational experience. However, their operational currency diminishes over time. Second, some civilian government employees (interagency) are detailed by their departments and agencies to teach for several years. As with military officers, they may have relevant operational experience and leadership skills but they may or may not be educators or have relevant advanced degrees. Third, the schools hire civilian academics who may have no military experience or exposure. Some may have teaching experience but may be more oriented to theory and research than curriculum development and mentoring. Finally, there are some retired government civilians who are valuable members of the faculty and staff for the interagency perspective that they bring to the institution. However, it is difficult to recruit some retired government civilians like Foreign Service Officers (FSOs) because in most cases, unlike military retirees, they must give up their government annuities in order to accept a government salary. Retired FSOs are in high demand on PME faculties, but have been unable to obtain waivers from the State Department.[611]

One quality issue that government civilians and military faculty members have in common is whether they were selected for PME in-residence or promotion. The Skelton Report suggested that it was generally inappropriate to hire military officers as faculty who were not competitively selected for attendance at in-residence PME or selected for promotion. The Panel also thought it inappropriate for faculty to teach the very next year following PME graduation without an intervening operational assignment during which officers could put their education into practice. Twenty years later both of these situations still occur.[612] The PME schools report that all military faculty members have completed the appropriate level of PME, but they may have done so by distance learning.[613] Those who attend PME through distance learning are not usually regarded as highly as those who are selected for in-residence education, except perhaps in the Navy. The Navy, unlike the other services, does not use competitive boards to select officers for senior school attendance.[614]

Beyond the appropriate level of PME completion, it may be particularly important for military faculty teaching at Joint Forces Staff College to be Joint Qualified Officers (JQOs) because of that school's singular focus on operational art and campaign planning.[615] All JQOs have completed JPME I and II and held a previous joint assignment. Joint and Combined Warfighting School military faculty is 68 percent JQOs (almost all the others will be JQOs by the time they finish their assignment) and the Joint Advanced Warfighting School has 100 percent JQOs as military faculty.[616]

Observation: It is generally not a good practice for those who were not selected for PME in- residence or promotion to teach those who have been selected. However, there are cases in which the specific expertise of the individual is appropriate for the teaching assignment.

Military Faculty – Operators, Specialists, and Academics

How do PME institutions attract top military officers as faculty? On the one hand, attracting top "performers" is difficult because these officers generally want to serve in leadership and command positions and are sought after by their services and the joint community. On the other hand, some schools have a reputation for attracting top officers for their faculties despite these considerations.

The service academies screen candidate faculty members quite rigorously for service as role models. In addition, these schools sponsor many of their military faculty members to obtain master's or doctoral degrees at top universities with a "payback" teaching assignment either immediately after graduate school or after an intervening operational assignment. Those who choose to return to operations after teaching tend to do quite well in command selection and promotions to O-4 and O-5 and those who choose to remain in academe can usually have full military careers of 20 or more years.[617]

The higher-level PME institutions offer mixed reviews of military faculty. Some are top leaders and teachers. Others are top leaders but have little teaching talent. Finally, some are great teachers and tops in their academic fields but would not be selected for in-residence PME (as top officers in their promotion year groups) and likely will not be promoted with their peers to O-6. Of all the primary-, intermediate-, and senior-level PME schools, MCU stood out. MCU has been willing to trade short faculty assignments for high-quality officers. The University selects highly competitive Marine officers but often loses them to promotion and command before a three-year tour is complete.[618] Student and leader feedback indicated MCU's teaching quality does not suffer under this approach.[619]

The services each set criteria for quality military faculty differently. Command and graduate degrees are the basic discriminators. According to Army policy-makers, "It is to the benefit of our Army, the services, and the officer students to provide the best quality officers as our instructors."[620] Air University works closely with the Air Force Personnel Center and those who manage colonels' assignments to "ensure highly qualified faculty members are assigned." In 2007, the Air Force started programs to competitively select officers to instruct at Squadron Officer School for two years and then attend Air Command and Staff College (ACSC), or to teach at ACSC for two years and then attend Air War College.[621] Naval War College seeks military faculty with current and relevant experience. Navy faculty billets are filled 46 percent by post-command commanders and captains or post-major-command captains. According to Navy testimony, "these are the Navy's best and brightest."[622] Intermediate schools want their instructors to have previously held O-5 level command, and the Command and General Staff College used to be able to boast that it had former brigade commanders as department directors. Operational tempo makes this much less likely today. The senior schools would like their military faculty members to have held O-6 command.

As for graduate degrees, PME schools report it is not always possible to find candidates who have in-residence degrees from high-quality civilian schools. Faculty at the service academies typically have a minimum of a master's degree and those without can be sponsored to go to graduate school full-time. The academies must have a certain number of Ph.D.s to maintain their Bachelor of Science degree accreditation. The Naval Academy (USNA) is slightly different in that it will allow officers who do not yet have a master's to work on their degree and teach at the same time. While the academies sponsor officers to graduate school, among the intermediate and senior schools Air University is the only institution that sponsors faculty candidates for relevant graduate degrees to keep new military faculty rotating into assignments there.[623]

For those military faculty members who are not yet JQOs, a career incentive for senior-level PME faculty is the potential for joint duty credit, which may benefit them in competing for future key assignments and promotions.[624] For those senior faculty and staff who choose to remain competitive, joint duty is key, and they might be less likely to volunteer for a faculty assignment if it did not carry the possibility of granting joint credit. By law, all of the joint school faculty positions (except the dean's at National even though the dean does teach) are eligible to be on the joint duty assignment list (JDAL).[625] Joint schools have faculty comprised of one third from each military department, while the service senior-level education (SLE) schools are required to maintain 60 percent of faculty from the host service and 20 percent from each of the others. Even though faculty members at service war colleges are also eligible by law because they teach JPME II, departmental policy does not grant JDAL eligibility to host service faculty on the premise that they do not gain enough joint acculturation. Non-host faculty members, by virtue of working in another service's environment while teaching joint matters, are eligible for "joint credit." This legal and policy framework ensures that high quality officers who want to be competitive for flag rank are more likely to seek out joint and non-host faculty opportunities, or at least not be opposed to taking them, than would be the case if these positions did not offer joint credit.

Although it is not the role of the PME institutions to award JDAL credit for the sake of recruiting and retaining quality uniformed faculty, most of the services and schools testified that, with stiff competition for every services' best officers, having JDAL credit could contribute to higher quality non-host faculty. The senior schools have signed a memorandum of understanding (MOU) that delineates the types and quality of officers the schools require for non-host faculty.[626] However, the school leaders reported that they often only receive one nominee who the schools can either take or accept a gap in the position.[627]

In contrast to JPME II faculty, intermediate-level education (ILE) non-host faculty members are precluded by law from JDAL eligibility.[628] The same issues apply as discussed above, except that even more of the intermediate faculty, because these officers are usually more junior in rank and therefore have the prospect of staying in the military longer, would hope to stay competitive for promotion and command selection. The goal of the Joint Officer Management system should remain to grant joint credit only for positions that truly relate to "joint matters" as defined in law. However, the Subcommittee recognizes that if these positions gain JDAL recognition, it is easier to recruit the highest quality officers for faculty duty. Similarly, if the non-host services cannot meet their obligation to send the highest quality officers to PME faculty duty, the institutions are faced with the choice of taking an officer of lesser quality or being short of non-host service faculty members. In addition, non-host service positions are filled, by policy, at a lower percentage than JDAL positions.[629] The inability to fill the non-host faculty billets could have the effect of causing the school to be out of compliance with the OPMEP, which would result in losing JPME accreditation. Although loss of OPMEP accreditation for this reason may only be an issue at Command and General Staff College right now because of the Army's decision to expand that course to 50 percent of all Army majors, this situation warrants continuing oversight. PAJE reports have already cited several other schools for a lack of non-host faculty. A memorandum of understanding among the intermediate-level schools is also being discussed, but it may not change the fact that the services likely will offer only a limited number of non-host service officers to the schools to choose as faculty.[630] Joint eligibility could be a tipping factor for attracting higher quality faculty.

As a result of the Subcommittee's preliminary PME work, including interviews with faculty and the testimony of leaders of the PME institutions, the House Report for the FY 2010 NDAA included an item of special interest on this subject.[631]

> The committee is encouraged to find that many military officers serving on the faculties of joint professional military education (JPME) phase II institutions are being appropriately recognized for their joint experience on the joint duty assignment list (JDAL). However, the committee finds that some of these officers, who should have this assignment designation, still do not. The committee encourages these institutions to be proactive in ensuring that military faculty at the senior level institutions receive the appropriate joint credit, and that those positions are included on the JDAL through the validation process. As part of its comprehensive review of professional military education, the committee is reviewing the faculty positions at JPME I level institutions. The Department of Defense should also review the faculty positions at the JPME I level institutions for JDAL suitability and provide the committee the results of its review.[632]

This position was also reflected in the FY 2010 NDAA Conference Report,

> The conferees believe that the limitation contained in section 668(b)(1)(B) of title 10, United States Code, that excludes assignments as instructor at joint professional military education Phase I courses from the joint duty assignment list may be inappropriate and could negatively impact the quality of instructors. The conferees intend to address this issue in the National Defense Authorization Act for Fiscal Year 2011 with the goal of improving instructor quality.[633]

***Finding:* Current law allows the services to advocate for host faculty billets at senior service schools to be included on the joint duty assignment list (JDAL). DOD and CJCS policy would subject those service requests to the usual two-part test for JDAL validation including involvement in joint matters (teaching joint subject matter) and acculturation. Before 2007, the service intermediate-level education (ILE) institutions were submitting their non-host faculty billets for JDAL validation. The services and CJCS routinely validated them. Changes in legislation in 2007 and implemented in 2009 resulted in ILE military non-host faculty billets being precluded from inclusion on, or going through the validation process to be on, the JDAL. This change can impact a school's faculty recruiting, its ability to meet ILE OPMEP accreditation standards, and the quality of its military faculty. Additionally, if faculty positions can qualify for JDAL status, this can assist faculty members in their career progression.**

***Recommendation*: Congress should amend the 2007 legislation so as not to preclude the possibility of JDAL credit being validated for non-host faculty at JPME I- granting institutions through the joint staff process as stated in U.S.C. Title 10 sec. 668 (b)(1)(B).**

In the discussion of joint assignment credit related to the quality of military faculty, the Subcommittee recognizes that through their own actions, the services' leaders can work to persuade officers that faculty assignments can offer viable paths to promotion and a full career. For instance, during the Subcommittee's hearing with the academy and primary PME school leaders, the Chairman had an exchange with the commander of the Air Force's Squadron Officer College. The Subcommittee learned that many Squadron Officer College faculty members were selected as non- volunteers because they perceived that a teaching assignment would not be beneficial.[634] This speaks to the esteem in which the services and joint community hold educational and professional development assignments. Joint and service leaders must be convinced of the positive value, as well as the real-world impact, of these assignments in order for their officers to be convinced. Additionally, the Service Secretaries can give direction to promotion boards and create other incentives for outstanding officers to volunteer for faculty duty.[635] The Subcommittee has an interest in civilian and military leaders of all ranks taking such a stance and taking action to increase the standing of faculty positions at all PME institutions for enhancing the quality of the faculty and promotion opportunities as well as key assignment and command selection for these officers.

The general question of the promotability of military PME faculty has at least two sides. Obviously, the higher the quality of faculty officers, the greater their chances of promotion. Competitiveness for promotion, in some cases, means officers will spend less time in their teaching assignments as they are selected for promotion or key assignments. On the other hand, maintaining continuity often dictates longer assignments. In fact, senior staff at USMA said they would prefer not to have an officer if he or she will not spend at least three years on faculty.[636]The Marine Corps has opted for accepting the most competitive officers for faculty duty, knowing that they may have to settle for shorter teaching assignments. While some officers want to stay competitive with their peers, others may "opt" to concentrate on an academic or educational path with repeated or lengthy faculty assignments. These officers know that the longer they stay away from operational assignments, the less competitive for command and promotion they may become.

As a result of these assignment dynamics, almost everyone the Subcommittee interviewed could name the few officers who have earned senior rank despite pursuing non-traditional career paths, gaining operational expertise, command assignments, attending PME in-residence, gaining advanced civilian degrees (even Ph.D.s), or serving as faculty members including General David Petraeus, Admiral James Stavridis, Admiral Patrick Walsh, and Brigadier General H.R. McMaster.[637]

Finding: **Rather than being products of deliberate personnel, joint officer management, and JPME systems, some well-known officers reportedly took personal initiative and risks with their careers. The Subcommittee finds that there should be a greater possibility for a few officers to follow a similar path, if qualified, which would include both intellectual and operational opportunities on the way to senior command.**

Civilian Faculty – Agencies, Academics, Contractors, and Former Military Officers[638]

U.S. Code Title 10 is now the predominant means of hiring civilian faculty for PME institutions. This section of the report addresses how the quality of civilian faculty is defined in the PME environment; how PME institutions seek to attract high-quality civilians; how a quality faculty is sustained; and, how quality is measured through faculty appraisal. Three other issues relate to the civilian faculties including the expansion of Title 10 authority to additional courses, the hiring of civilians considered "pure" academicians, and the hiring of contract instructors. In general, the Subcommittee found a wide variety in the application of Title 10 hiring authority in the PME enterprise. In fact, the differences are so vast that one former NDU president suggested that at least some standardization would be helpful to the various school and service leaders to reduce perceptions of arbitrariness so that no institution would suffer by comparison to its peers.[639]

Defining Quality

The PME institutions report variations on ways to judge the quality of their civilian faculty. Some want civilians who are nationally recognized in a relevant field.[640] Others want teaching experience and a record of publishing in peer-reviewed journals.[641] At the intermediate-level a general knowledge of adult education methodology and a passion for teaching was cited as being important.[642] Others defined quality by position. For tactics instructors, a master's degree might suffice, but for others a Ph.D. is required.[643] Finally, some schools highlighted that they judge quality, in part, by the potential a faculty member has for making significant contributions to the school.[644] At the senior level, Ph.D.s or terminal degrees (J.D.) were valued even more. Some schools are more apt to hire retired military officers who are Ph.D. candidates but have not completed their dissertations.[645]

Attracting Quality

Some of the PME schools, particularly those (other than the service academies) outside of the National Capital area, report challenges in attracting top-tier civilian faculty when competing for the same talent as prestigious civilian universities.[646] In this regard, money matters. Dr. Williamson Murray testified to the Subcommittee that the PME system is "seriously underfunded" and that the schools need increases in funding to ensure the quality of the faculty.[647]

At the pre-commissioning-level, salary was only one consideration. The USAFA dean testified that other attractions include an outstanding student body, undergraduate research, an important mission, and a desirable location. She mentioned short "contract length" as a detractor.[648] Naval Academy leaders cited as their greatest incentives a balance of teaching with scholarship opportunities, funding for professional development, outstanding students, tenure after six years of outstanding performance, and competitive starting salaries. The commandant reported that detractors include lack of long-term pay parity, faculty housing, and tuition assistance for family members.[649]

All the intermediate schools cited pay and benefits as incentives. Many also consider their outstanding students as an attraction. In addition, some cited degree-granting authority and regional accreditation, along with support for research travel. ACSC leaders consider detractors to be their location in Montgomery, Alabama and lack of tenure combined with the "vagaries of faculty management policy changes" because of rapid turnover of senior leaders and issues of "academic freedom."[650] CGSC reported detractors such as lack of tenure and copyright as well as high classroom teaching loads that negatively impact time for research.[651] Marine C&SC leaders consider their location near the national Capital and quality of other faculty as incentives for new faculty, while the lack of tenure is a detractor.[652] Joint Forces Staff College cited academic freedom and rewards for exceptional performers on the positive side. On the negative side, the small number of faculty members and gaps in military faculty positions do not allow civilian faculty sufficient time for research and publishing.[653] NWC seconded many others' significant incentives but also reported that senior faculty pay is not competitive, teaching and curriculum development leaves little time for research, and retired State Department personnel forfeit their annuity.[654]

The absence of tenure is one of the most often cited impediments to attracting highly qualified civilian faculty. The Naval Academy is the sole PME institution with a civilian tenure system.[655] In fact, when asked "what's the most important element to hiring quality faculty," the dean responded in no uncertain terms that a school cannot have a quality faculty without a tenure system.[656] Other exceptions to the rule include the NWC's several positions that are "without term," which are civilian positions of indefinite length, and there are a number of Title 5, or General Schedule (GS), faculty members remaining at some schools.[657] Title 5 employees inherently have more job security.[658]

Tenure is not an absolute necessity as long as other conditions of employment balance any perceived disadvantage. As a former president of NDU observed, conditions of employment are not standardized among PME schools.[659] PME schools hire civilian academics under authorities granted under Title 10 of the U.S. Code. Title 10 simply allows the Secretary of Defense and the Service Secretaries to hire the number of civilian faculty required as "professors, instructors, and lecturers" for 10-month PME courses and the 10-week Joint and Combined Warfighting School course.[660] The service academies are treated separately. The Department has used this flexibility to hire high-quality faculty with appropriate credentials at a higher pay rate than the GS system and, at the same time, offer the schools flexibility to not retain faculty members on a more permanent basis in case of program changes. Generally, Title 10 faculty can be hired quickly through a traditional academic search committee procedure, or as reported to the Subcommittee, "by word of mouth."[661] Although not dictated by law, the PME institutions usually offer new faculty a one- or two-year appointment and then renewal for three years for satisfactory performance. By policy, PME schools can offer five- to six-year renewals if they choose to do so. While most civilian faculty members seem to be renewed as long as they perform satisfactorily, appointments can be non-renewed if a program is terminated or if the school's curriculum moves in a direction that the employees' specialty will not support.[662] Transparency in this process is important to faculty members.[663]

The Subcommittee learned that the various PME schools do not use written "contracts." Instead, the schools typically issue appointment letters to their Title 10 faculty. Some schools, including the Naval War College, issue very detailed appointment letters in which expectations for faculty members are well-defined, while others issue very brief letters.[664]

Observation: Naval War College (NWC) uses detailed appointment letters and the Subcommittee suggests that other schools could use the NWC example as an instructive model.

In addition to appointment letters, all of the schools rely on their faculty handbooks as well as directives, policy letters, regulations, service instructions, and faculty operating instructions to communicate expectations. According to these documents, any appointment can be terminated for cause or loss of a required security clearance.

Maintaining Quality

For renewal, most faculty directives say staff will be judged on performance in the areas of teaching, research, and service. However, some school policies make it clear that there is no requirement for the administration to give reasons for non-renewal, even in the face of documented outstanding performance. This is particularly true for NDU's faculty policies.[665] On the other hand, NDU leaders stated that despite reports of confusion or unfairness in renewal policy and practice, faculty members are treated well. Morale and sustainment of an outstanding faculty, however, may depend on even one incident in which any administration is perceived to be arbitrary.

The renewal process for civilian faculty appointments was also characterized by a lack of timeliness at two institutions, National and Air War College. At National renewals must be processed through the University's Human Resources Directorate (HRD), then the NDU chain of command, and then the Defense Financial and Accounting Service. Although ICAF faculty and staff had no complaints, National's leaders admitted that significant personnel turnover and vacancies in HRD caused delays.[666] As stated previously, the Air War College reported that their renewal process is not timely because of the multiple layers required for approval at the college, the Spaatz Center, and Air University.[667]

The renewal process, like all other Title 10 activities, varies by service and school except that regardless of the length of the first appointment, the first year for new faculty is generally probationary. Some leaders and faculty members suggested that, to be effective, a two-year probation period is really more reasonable. Reappointment is usually for one to five years based on performance with three being the standard.[668] The service PME institutions believe that their processes are well-documented and very transparent.[669] The Subcommittee heard no complaints from NWC or MCU faculty on renewal policies. On the other hand, AWC reported that some faculty members find it disconcerting that the final approval for renewal rests with the AU commander instead of their school commandant who knows their work best.[670]

JFSC reports that NDU's president has the sole authority for renewal decisions, based on commandants' recommendations according to *NDU Regulation 690-4*. NDU policy specifies that no faculty member is entitled to renewal and that non-renewal does not constitute involuntary termination. However, renewal is based on "demonstrated exceptional professional experience and competence." Faculty members are expected to remain current in their field and sustain their performance. They are notified in writing as early as possible if they will not be renewed.[671]

At the NDU senior school level, policy is internally inconsistent as to when faculty will be notified of non-renewal decisions. *NDU Regulation 690-4* states "employees should receive final official notification at least eight months before their current employment term ends." But a non- renewal recommendation only needs to be forwarded to the president of NDU with six months remaining on an appointment. HRD notifies the school of approval of renewal only 30 days prior to the effective date. In addition to confusing timelines, the policy regulation for a college peer review panel, after which the commandant's and the NDU provost's recommendations are forwarded to the NDU president. NDU faculty reported that a peer review is not always formally included in the process and renewal recommendations are often based instead on advice from the department head, dean, or chief of staff to the commandant. While National's commandant reportedly does not cite reasons for non-renewal in counseling terminated employees, ICAF leaders reported that performance is definitely a factor and is discussed with employees.[672]

Other appointment variations include one-year limited appointments and adjuncts. West Point and USAFA are using one year appointments to fill gaps for uniformed members who are deployed. These are non-renewable, but civilians hired for these positions can compete for permanent openings if they become available.[673] ICAF and other schools hire adjunct faculty for the potential savings and to hire specialists to teach electives.[674] The Subcommittee understands this may be appropriate for some specialized courses such as language classes, but acknowledges that many faculty members believe that one-year appointees and adjunct faculty may not have a continuing commitment or attachment to the institution, other faculty, and the unique PME mission. Adjuncts and single-year hires usually do not perform the same service and mentoring responsibilities, leaving more duties to the longer-term faculty and staff. The PAJE team has also called attention to the fact that "full time equivalents" are problematic when it comes to assessing faculty ratios and student-to-faculty ratios for accreditation purposes.[675]

Observation: The Subcommittee recognizes that no institution has an unlimited budget, but continued reliance on short-term faculty should not be used as a regular tool without an assessment of the long-term impact on the institution and other faculty.

Measuring Quality and Title 10

PME school leaders cite their faculty handbooks as setting out fair and transparent appraisal and renewal policies. These include student evaluations and, in some cases, peer evaluations. The handbooks also show that they have appeal processes for non-renewal or problematic appraisals. In fact, school leaders suggest that because very few faculty members are non-renewed, they have *de facto* tenure systems. However, it only takes one perceived arbitrary non-renewal to have a chilling effect on faculty morale and quality recruitment and retention.[676]

The Subcommittee heard a variety of perspectives on the issues of appraisals and renewals. Many Army War College faculty members said they like their renewal system because they can renegotiate their salaries more often.[677] USAFA has reported trying a new rotating appointment system to improve faculty members' sense of employment stability; the results of this change are not yet available.[678] West Point leaders and faculty have stated that their system of renewal might be as good as tenure, if not better, because civilian university tenure systems use the same criteria and time lines for renewal decisions.[679] The Naval Academy, by comparison, has a tenure system under which they award new faculty an initial appointment of not more than three-years. New faculty members in possession of an earned doctorate enter as assistant professors. They then have five years before they can be considered for promotion to associate professor. Tenure normally comes with promotion to associate professor. Civilian faculty members normally have a maximum of six years to be promoted to associate professor and earn tenure.[680] The 2004 Larson Report (by retired Admiral Charles Larson) on permanent-military professors at USAFA, tasked by Congress in 2003, stated that while the Naval Academy tenure model did not have to be a "one-size-fits-all" for the academies, the absence of a tenure system would place limits on civilian faculty as time went on. Limiting factors include temporary employment status, lack of department head or other leadership positions, lack of a role in high-level curriculum decisions, limits on professional development, and holding the percentage of civilian faculty down.[681]

Faculty members at a number of schools reported discomfort with the Title 10 system in general. Some think the policies and practices governing the balance of power between administrations and individual faculty members tilt too much toward the former. Faculty at ICAF and National cited concerns about arbitrariness if no reason for non-renewal or lengths of reappointments must be given in the face of otherwise outstanding measures of performance.[682] Faculty at Army War College, ICAF, National, USMA, and USAFA cited concerns about a level of trust if power is perceived to be one-sided.[683] This challenge is compounded by the fact that most PME schools do not have independent faculty senates or other representative bodies that can take faculty concerns to the superintendent without first being screened by the dean or commandant.[684]

Complaints about various schools' systems that reportedly would have been voiced through faculty senates included concerns about job security and competitiveness after PME employment, appraisal transparency and consistency, stated policies not being followed, and academic freedom. Some faculty members at National, ICAF, USAFA, and USMA cited concerns about competing in the civilian university system after working in a non-tenured position, particularly if they are let go without explanation.[685] Some National and ICAF faculty members claimed that their appraisal system lacks transparency and consistency and that the procedures in the faculty handbook and school policies are only loosely followed, if at all.[686] Finally, USMA, USAFA, ICAF, National, and Air University faculty members cited concerns about academic freedom; since they can be denied renewal without being told why, faculty members fear retribution if their scholarship is controversial.[687] At civilian universities, those who have tenure are thought to have the most academic freedom.

Some PME leaders and faculty members said that they would prefer tenure systems for almost the opposite reason as others embrace it. Instead of believing that tenure tilts the balance of power between the administration and individual faculty members more to faculty, they say it holds the faculty to higher standards and guarantees better performance. These faculty members and leaders think that if faculty members had to go through major, university-wide, tenure reviews at set intervals there would be greater incentive for better performance.[688] Other faculty members do not like tenure systems because they see them as discriminatory against those who focus on teaching rather than publishing and against faculty members with family obligations that would compete with a rigid tenure schedule.[689]

***Finding*: Some civilian PME faculty members think that they do not have timely, fair, and transparent renewal and appraisal systems. Perceptions on these issues matter for recruiting and sustaining quality faculty. The Naval War College and Marine Corps University provide positive examples. While allowing for some flexibility among the services' unique programs, more consistent DOD policy in this area is needed.**

***Recommendation*: The Department and the services should reexamine the nature of the employment relationship with civilian faculty hired under Title 10 authority and consider adding to or enhancing existing policies that would address transparency and stability issues while retaining institutional flexibility to hire quality civilian instructors without unnecessary delays.**

Another issue that arose regarding to civilian faculty was the requirement to balance stability with requirements for "new thinking" or new areas of emphasis. All of the PME institutions require that civilian professors keep current in their discipline while emphasizing teaching. Operational or policy currency is sought from agency and military faculty through continual one- to three-year rotations. At the more senior levels of PME, the balance between stability and new thinking brings another requirement. The schools must balance the need for seasoning and experience among professionals who can hold their own with mature professional students with the need for the energy and innovation usually associated with younger faculty. For instance, the Marine Corps War College (MCWAR) cited younger professors as having "academic vitality" that keeps the curriculum "current and vibrant" and brings in new teaching methodology and technology.[690] Dr. Andrew Krepinevich went further, asking whether changes in PME curriculum had engendered turnover in faculties or "plainly speaking, has the 'legacy' faculty been shoehorned into a very different curriculum?"[691] The Army and Air War Colleges agreed that faculty experience and seasoning were helpful with their unique student bodies, but that younger professors who could hold their own in seminars were desired as well. They pointed out, though, that with few positions and low attrition, turnover occurs slowly. None of these institutions intended to push out older professors, but NWC did cite a concern about the "graying" of their faculty if younger faculty were not brought in.[692] The Subcommittee heard several allegations of "ageism," including the example of the non-renewal of, or creating a hostile work environment for, senior faculty who kept current in their discipline and received outstanding performance reviews and student and peer critiques.[693]

Some experts, including retired Major General Robert Scales, also believe that foreign nationals should not be restricted from the faculty at the senior colleges because of security clearance requirements.[694] Scales and others noted that current law restricts foreign participation, but that foreign faculty are often "experts in service school areas of study and can bring helpful, international perspectives to student learning."[695] Many North Atlantic Treaty Organization (NATO) partners actually hold sufficient security clearances, but they are barred from faculty duty.

Finally and perhaps most importantly, commanders, commandants, and presidents of some PME institutions cited concerns that they could not easily recruit retired Foreign Service Officers and former ambassadors (as well as several other categories of civilian employees) as instructors because these individuals have to obtain Office of Personnel Management or State Department waivers or give up their federal annuities in order to take a PME position.[696] The FY 2010 NDAA encourages agencies, including the Department of State, to use existing authorities to allow the PME institutions to hire qualified annuitants under similar conditions to those for retired military officers.[697]

***Finding*: The quality of civilian faculty could be increased with regard to the interagency perspective by allowing retired Foreign Service Officers and Ambassadors to join PME faculties without giving up their federal retirement pay. These civilians may join the faculties now only by giving up their annuity or with a waiver. The State Department, in particular, has not granted waivers.**

***Recommendation*: The State Department and other government agencies with employees not under one of the excepted retirement systems (Civil Service Retirement**

System and Federal Employees Retirement System) should grant waivers based on FY 2010 NDAA, sec. 1102.

Three other significant civilian faculty issues beyond defining, attracting, maintaining, and measuring quality surfaced during this study: the expansion of Title 10 authority to more staff at PME institutions, the definition of "pure" civilian academics and civilians in leadership positions at the academies, and contractors as military faculty substitutes at CGSC.

Title 10 Expansion

Title 10 of the U.S. Code gives the Service Secretaries the authority to hire "as many civilians as professors, instructors, and lecturers" as they consider necessary. However it also states that "this section shall not apply with respect to professors, instructors, and lecturers. ...if the duration of the principal course of instruction offered at that school is less than 10 months."[698] Although FY 2010 NDAA section 1113(d) allows the Department broad hiring flexibility, Marine Corps University (MCU), Command and General Staff College, and Air University have requested expanded Title 10 hiring authority.[699]

The Marine Corps wants to use Title 10 authority to hire faculty for courses that last less than 10 months. Specifically, this would be used for hiring enlisted PME course faculty.[700] MCU equates the kind of civilian faculty necessary for these courses to Ph.D. professors hired for the intermediate and senior officer courses with salaries and flexibility as the two key reasons for the authority. In addition, Title 10 allows for rapid adaptation of curricular content to reflect changes mandated from lessons learned in ongoing conflicts. These courses include six Staff Non-Commissioned Officer (NCO) Academies and the School of Marine Air Ground Task Force Logistics. MCU states "expansion of Title 10 authority to hire qualified, professional educators as faculty with specialized skills and disciplines is essential to foster the implementation of adult learning theory and educational technology in delivery and student assessment."[701] MCU argues staff NCO academies rely primarily on active-duty Marines who are fulfilling positions outside of their primary military specialty area. They are not professional educators and do not possess the pedagogical expertise to implement current adult learning theory using the newest educational technology. Within Title 5, there is no specialty series identified as "professional educator." Title 10 would allow MCU to hire through faculty searches rather than the Title 5 process which requires selection from a list of minimally qualified employees from an applicable GS series, which again, does not exist. Contractors, similarly, are not an optimal solution as the Department has mandated significant reductions in their use.[702]

The Air Force asked the Subcommittee to amend Title 10 to: (1) hire "faculty" rather than the more the specific "professors, instructors, and lecturers" to include researchers; and, (2) allow for courses of shorter length. The Air Force has not yet addressed the specifics of why those employed by "the Barnes Center, the Spaatz Center, the Holms Center, Air Force Research Institute, the Muir S. Fairchild Research Institute, or the Eaker College for Professional Development" should not be hired under existing authorities.[703]

Civilian hiring authorities extend to a wider group of military education centers beyond the intermediate and senior PME institutions including Defense Acquisition University, Naval Postgraduate School, the Air Force Institute of Technology, and the service academies. The same description of those covered, "professors, instructors, and lecturers," is included for all, which is relatively standard under Title 5 of the U.S. Code in describing the population exempt from normal civil service hiring procedures. The Subcommittee did not study all of the institutions covered by these statutes. Jurisdiction remains with the House Government Oversight and Reform Committee for DOD schools for military family members. For these schools, Title 5 includes the exempted category of "teacher" (as defined under 20 U.S.C., section 901). Title 5 allows the Secretary of Defense to define teaching positions by regulation. However, Title 10 does not allow the Secretary to define what is meant by "professors, instructors, and lecturers" by regulation.[704]

***Finding*: Title 10 sections 1595, 4021, 7478, and 9021 restrict the hiring of civilians both by course length (10 months) and position description ("professors, instructors, and lecturers") at some PME institutions.**

***Recommendation*: If FY 2010 NDAA section 1113(d) is insufficient to attract quality faculty, Title 10 sections should be amended to allow flexibility to hire civilian academics for PME institutions at a competitive salary and with flexible terms of service, while allowing for the reasonable ability to hire the appropriate personnel.**

Civilians in Service Academy Governance and "Pure" Civilian Faculty at the USAFA

The Naval Academy has a long history of employing an evenly balanced civilian and military faculty. In contrast, USMA and USAFA had almost exclusively military faculty until Congress provided Title 10 hiring authority to the Air Force and Army to "extend to the Air Force Academy and the U.S. Military Academy the same flexibility for hiring civilian faculty that exists for the Naval Academy."[705] The legislation, as amended in 1993, gives the Secretary of the Air Force the discretion to hire as many civilians for USAFA as the Secretary considers necessary.[706] Faculty members raised the issue of whether USAFA hires enough "pure" civilian academics with the Subcommittee. This issue was previously described in the 2004 Larson Report.[707]

A related issue is whether civilians have a voice in the governance of PME institutions. Faculty members at USMA and USAFA, among other PME institutions, stated that they did not believe civilian faculty had enough of a voice in school governance.[708] Part of the issue at the two academies, in addition to not having faculty senates, is barring civilian faculty members from competing for department head and dean's office senior positions.[709] The 2004 Larson Report concluded that "A Maturing Civilian Faculty Will Increase Pressure on the [Air Force] Academy to Provide them a Greater Role," stating "as the civilian faculty has matured and gotten more senior in academic rank, it is inevitable that pressures will increase for tenure, an expanded leadership role and a greater role in curriculum development outside individual departments."[710] In response to the Subcommittee's questions, both schools' leaders pointed to a number of civilians who serve on the deans' staffs.[711] Civilian faculty members, in response, pointed out that almost every one of these senior civilians is a retired military officer, almost all were graduates of the institution, and almost all had spent a considerable amount of time as uniformed faculty members at those two schools.[712] The Larson Report highlighted several problems that could arise, and the Subcommittee found them to be still relevant at USAFA in 2009. These included: desire for department and administrative leadership positions; "contingent employment" under contracts versus tenure; participation in academy curricula decision-making; limits on professional development particularly in research and publication; and minority representation as civilians among a majority military faculty.[713]

The governance issue indicated an even more specific concern that emerged both during the Larson study and the Subcommittee's review of PME. At USAFA, a significant number of civilian faculty members cannot be described as "pure civilians," as they can at the Naval Academy and West Point. The USAFA dean reported 30 percent of civilian faculty members are retired military, including some who do not have Ph.D.s.[714] The concern here rests on the benefits of employing civilian faculty at the academies, rather than on the optimal percentage of civilians.[715] Hiring civilians can bring in highly-qualified professional educators and academics in order to increase the number of faculty with Ph.D.s who have more standard academic credentials; to increase the diversity of faculty to which the students would be exposed; to provide students, military faculty, and staff an outside perspective; to enhance civil-military relations; and to improve students' critical thinking abilities.[716] According to the Larson Report, civilian professors add "a fresh and often provocative world view not bounded by military culture," "doctoral-level currency and depth in their academic disciplines," "access to powerful educational, professional, alumni and research networks," and "opportunities for cadets and midshipmen to build awareness of the increasingly large and critical roles played by civilians throughout the Department of Defense and their parent service."[717] The Larson Report's recommendation on "Hiring Civilian Faculty" said, "to ensure the maximum strength of the civilian element of the faculty, future civilian hires should be 'pure academicians' from civilian higher education."[718]

The dean's testimony to the Subcommittee indicated that USAFA does not believe the question of "pure" civilians is an issue.[719] The dean reported that "…about one third of civilian faculty are retired military. ...And we try not to, in our processes, advantage military or disadvantage military, retired military."[720] The Air Force Academy, in response to questions after the hearing, cited a civil case, under which the institution was sued by a former military member who was not allowed to compete fairly for a civilian faculty position.[721]

> However, specific application of this recommendation is constrained by other legislation such as the Uniformed Services Employment and Reemployment Rights Act of 1994 (USERRA, 38 U.S.C. § 4301- 4335). USERRA is a federal law intended to ensure that persons who serve or have served in the Armed Forces, Reserves, National Guard or other "uniformed services:" (1) are not disadvantaged in their civilian careers because of their service; (2) are promptly reemployed in their civilian jobs upon return from duty; and (3) ***are not discriminated against in employment based on past, present, or future military service*** *(emphasis added)*. …Such a person shall not be denied initial employment, reemployment, retention in employment, promotion, or any benefit of employment by an employer on the basis of that…performance of service.
>
> Previous USAFA attempts to hire "pure academicians" instead of equally or more qualified military retirees resulted in a complaint to and an investigation by the Office of Special Counsel (OSC). As a result of the investigation, OSC informed USAFA that they would bring an action before the Merit Systems Protection Board unless USAFA strictly complied with the anti-discrimination provisions of USERRA. After a discussion with OSC and a review of the law, USAFA thereafter hired the soon-to-be retired military applicant for a civilian faculty position at USAFA.[722]

The resolution of the issue of hiring pure civilian academics may actually hinge on requirements specified for any particular faculty position, not on whether military retirees are "more qualified" for positions as USAFA asserts and was found in this case. Instead, hiring entails fine lines drawn between those who are "equally qualified." The Subcommittee does not expect a more qualified applicant to be disadvantaged. Nonetheless, USMA and USNA do not seem to have the same challenges complying with the Uniformed Services Employment and Reemployment Rights Act of 1994, Office of Special Counsel, or the Merit Systems Protection Board and supporting "pure" civilian representation on the faculty.[723]

Despite the lack of specific ratios of civilian to military faculty directed in law, the reported assumption at the time of the Larson Report and from a number of current civilian faculty members, was and is that constructive change and transformation seldom comes from within an organization or from those invested in a system that they lived within for 20-30 years. This situation is exacerbated at USAFA by the fact that civilian faculty cannot effectively hold senior positions and some feel that they do not have an effective voice in governance.[724]

***Finding*: USAFA has not done enough to hire pure civilian faculty. USAFA could consult the Larson Report recommendations in this regard.**

***Recommendation*: USAFA should endeavor to hire a greater number of "pure civilians" as faculty members.**

***Finding*: Some civilians at USAFA and USMA would like to compete for more senior academic leadership positions. There is value in having civilian academics in senior positions to ensure that all voices are heard in academic governance.**

***Recommendation*: USAFA should allow "pure civilians" to compete to serve in positions that have a more significant role in academic governance.**

Contractors with Operational Experience as Military Faculty Substitutes

Command General Staff College (CGSC) seems, at first, to have the opposite challenge from USAFA. Because of operational tempo and the increased requirement for faculty due to the universal ILE policy, CGSC has hired more civilians, particularly contractors, to replace unavailable military instructors. The balance is now about 70:30 in favor of civilians. Contract civilians rather than Title 10 appointees comprise the majority of those civilians. The Army would seek to bring this ratio back into better balance with a higher number of military faculty members who have recent operational experience.[725] Hiring contractors, who are recently retired military officers, seems to be an appropriate expedient, but unfortunately, they have a short-lived operational currency. After their first two to four years, they no longer fit the appropriate military faculty substitute profile. Army CGSC students reported to the Subcommittee on a lack of instructor currency. Students also identified quality problems with some of the contract faculty who are not hired individually by the school but as subcontractors by a private company. Some of these retired officers had not been selected for ILE in-residence, nor had they been promoted with their peers, giving the impression that they had not "made the grade" while their students had or would.[726]

Observation: Until the Army is able to correct the balance between military and civilian faculty, Command and General Staff College should better screen contract faculty members for quality and operational currency.

Civilian and Military Faculty Issues

Academic Freedom

Academic freedom is a key element in meeting degree accreditation requirements and in attracting quality civilian faculty. Although academic freedom relates mostly to discussions of civilian faculty without a tenure system, it also has wider application to all faculty members and the PME mission to foster critical thinking and oral and written communication. All the PME institutions subscribe to the principle of academic freedom even within a military context. The only differences countenanced in this distinct context are compliance with the Uniform Code of Military Justice and other laws, protection of classified information, and prevention of sensitive technology transfer. Academic Freedom is not a *carte blanche* for faculty members or students to say or write anything they desire without a related responsibility to their academic community. Their presentations and writing must be objective, professional, and within the scope of their expertise.

Academic Freedom is defined by the American Association of University Professors (AAUP) statement issued in 1940 with the interpretations of 1970. NDU accepts this construct in concert with Title 10 and DOD Directive 5230.09, *Clearance of DoD Information for Public Release*, in its policy statements and regulations. NDU defines academic freedom as:

> Freedom to pursue and express ideas, opinions, and issues germane to the University's stated mission, free of limitations, restraints, or coercion by the University or external environment. Academic freedom is the hallmark of an academic institution. We expect all members of the University community to understand the importance of and to practice responsible academic freedom.[727]

The question is not whether the PME institutions endorse academic freedom. They do. The issue is whether it exists in practice and in the perceptions of their faculty. For instance, NDU faculty members brought the DOD Directive to the Subcommittee's attention as a possible problem. Although papers written by faculty and students that are not intended for outside publication are exempt, the directive encompasses all academic papers meant for public presentation or publication.

> To ensure a climate of academic freedom and to encourage intellectual expression. ...Information intended for public release or made available in libraries to which the public has access shall be submitted for review. Clearance shall be granted if classified information is not disclosed, *DoD interests are not jeopardized, and the author accurately portrays official policy, even if the author takes issue with that policy.*[728]

Interpreting whether "DOD interests are jeopardized" can be problematic if a scholar takes issue with an official policy and does a credible job of doing so.

It was brought to the Subcommittee's attention by the leader of one of the PME institutions that recent controversies in civilian academe have led to a new AAUP Report, which points in the opposite direction from the DOD Directive:

> Response to help faculty with what they can and can't say in the classroom... especially on controversial or political issues. "We ought to learn from history that the

vitality of institutions of higher learning has been damaged far more by efforts to correct abuses of freedom than by those alleged abuses. …The essence of higher education does not lie in the passive transmission of knowledge but in the inculcation of a mature independence of mind."[729]

Bachelor's and master's degree accreditation at the PME institutions depends on endorsement of academic freedom policies and the fair implementation of them. For example, Middle States Accreditation Standards require "providing support for academic and intellectual freedom."[730] The question posed during its review was, "To what extent does [the school's] mission, values, goals and strategic planning process reflect a commitment to the principles of academic freedom and what evidence validates this conclusion?"[731] The Higher Learning Commission of the North Central Association of Colleges and Schools includes in its Assurance Section core component 4b under fulfillment of criteria, criterion four, assessing academic freedom under, "the exercise of intellectual inquiry" as integral to its educational programs.[732]

At PME institutions, experts recognize that there must be an appropriate balance between military and civilian academic cultures. In response to the Subcommittee's questions after the first PME hearing, Dr. Alexander Cochran answered, "The expectation (indeed the obligation) within the military culture to offer alternative views, particularly in the decision-making process, is strong." Dr. Janet Breslin-Smith added that atmosphere is as important as stated policy;

> When a leader in an academic institution suggests that certain speakers should not be invited, that administration policy should not be questioned, that certain schools focus too much on history and policy criticism, great harm is done to military officers. …The goal of senior officer education is critical analysis and strategic thought. …Faculty and students need to be free to question, to reconsider, to challenge.[733]

In practice, open inquiry and challenging authoritative views are weighed against ensuring classified information is not divulged and that the Uniform Code of Military Justice is not violated. The faculty's modeling of the ability to question intellectual authorities must be balanced by professionalism and good judgment. Jackson Niday and Kathleen Harrington, professors at USAFA, contend that academic freedom is linked to the development of critical thinking and that this is not only desired but imperative at service academies:

> In the end, the strongest argument for why military academies must acknowledge the importance of academic freedom lies in perhaps the most hallowed of military rites of passage. …In short, officers' oaths hold them more responsible to think as leaders—to think more critically, more accurately, and more independently. Preparing officers to think critically, accurately, and independently requires a certain kind of reflective space—space to challenge the status quo, space to make honest mistakes, space to learn from one's errors. Academic freedom provides that space.[734]

Army War College and Marine Corps University leaders recognize that academic freedom is ultimately important.[735] Air Command and Staff College leaders agree "Academic freedom is fundamental in producing students who are agile, critical thinkers capable of leveraging new ideas in the complex and fast-paced environment of military operations."[736] Nonetheless, the Subcommittee heard allegations that academic freedom has been violated at several schools.[737] Faculty members reported that some censorship was external or imposed by the institution, and some consisted of self-censorship. The publication and speaking review process was cited as a specific concern at Air University and at USAFA.[738]

Air University (AU) policy is based on the AAUP definition, however, some AU faculty members claimed they were prevented from responding to press interviews in a timely manner and that writing on controversial issues is a problem.[739] One AU faculty member reported, "We have academic freedom, except when we need it." Other faculty members agreed that it took so long for the University to approve their invitation to speak to radio and television reporters, that they often lost the chance to do so for lack of timeliness. Many of these individuals reported that they decline invitations without even asking for permission or that reporters stopped inviting them to comment.[740] The situation at AU can be contrasted with the Naval War College (NWC) position. NWC allows its faculty to notify the school before or after press inquiries, interviews, or giving expert advice. Faculty members know they can respond but must provide a written or verbal disclaimer.[741] Despite faculty criticism of the practice, AU similarly states in policy documents that their faculty should respond to the press in a timely manner.[742]

Within the Air Force, Air University and USAFA have different regulations and instructions. However, USAFA professors had similar concerns to those at AU with respect to a slow clearance process.[743] According to the 2008 USAFA Faculty Operating Instruction (FOI) 35-101, the clearance process and academic freedom are treated together; there is no mention of the AAUP guidelines in the FOI.[744] The process is often too slow to support faculty conference presentations, much less time sensitive press interviews.[745] Manuscripts and presentations go through the Director of Research to try to prevent classified material and technology transfer compromises for the sciences and engineering, but the process also catches literature and history manuscripts (but not fiction) in its sweep as well.[746]

As for USAFA's intellectual environment of inquiry and presentation, current and former faculty members reported a number of instances of slow clearance and lack of academic freedom verified by reporting and documentation. The Higher Learning Commission also recently documented this situation in USAFA's latest accreditation report. Faculty cited the refusal to give timely clearance for publication and presentation as a problem to both the Subcommittee and to the accreditation body.[747] Less public, but also of concern are USAFA faculty members' claims of self- censorship. [748] Together these instances foster perceptions of limited support for open inquiry and for academic freedom at USAFA. According to faculty interviews and the 2009 accreditation report, these episodes have had the effect of chilling the intellectual environment.[749]

Nonetheless, the Subcommittee is optimistic that USAFA's leaders will address these challenges having learned that the dean has convened two committees – one on faculty speakers and one on publications and speaking. The committees are chartered to look at the Air Force regulations and the current USAFA operating instructions and to make recommendations on how each should be revised.[750]

***Finding*: For publication and press interviews, USAFA policy diverges from those of other academies and PME institutions. Although FOI 35-101 supports academic freedom and "maximum clearance in minimum time," USAFA practices have created at least a perception of an environment that is not conducive to open inquiry in some subject areas.**

***Recommendation*: USAFA should continue to address the faculty's perception that USAFA does not fully support academic freedom.**

The situation at USAFA contrasted starkly with that at USMA. The few faculty members the Subcommittee met with did not seem to have concerns about the school's academic freedom policy or practice. On the other hand, a May 2009 USMA Academic Freedom Advisory Committee paper reported concerns and made four recommendations including that USMA reaffirm the 1940 Statement on Principles on Academic Freedom; strengthen due process and peer review of adverse action against professors; ensure faculty members have access to all websites and information; ensure faculty member schedule flexibility; ensure equal sabbaticals and independent selection of sabbatical work; and, ensure that offices outside the academic program (like the Judge Advocate General and public affairs officer) understand and support the mission of faculty members.[751] USMA reports having addressed these issues appropriately by revising the faculty manual in 2005 and offering "substantial protection" to associate professors and full professors including those protections related to academic freedom. The dean also reported looking into revising the language in faculty appointment letters to comport with the revisions to the manual. Finally, the dean noted that, given his actions, faculty members raised no concerns on these issues with the Middle States Accreditation team in September 2009.[752]

If PME institutions want to attract quality civilian faculty and allow military faculty to excel while modeling professional critical thinking for students, commitment to academic freedom will be the measure. Clearance for publication, presentation, or press interviews should be a timely process and recognize that social science and humanities work, while less likely to be classified, may be more likely to make people uncomfortable. Any limits on academic freedom should be clear in appointment letters. It might be time for all the PME schools to review their policies and practices.

***Finding*: The policies at the PME institutions on academic freedom vary. Some faculty members at several of the PME schools have been critical of their institutions' policies or the implementation of those policies. This is one of a number of factors that can impact the recruiting and retention of a high-quality civilian faculty.**

***Recommendation*: While the Department should review its policy on academic presentations and publishing, all of the PME schools should review their policies and practices on academic freedom. Institutions should recognize the AAUP 1940 Statement of Principles and the 1970 Interpretive Comments as the "gold standard" for academic institutions with appropriate variation for unique military requirements in the Uniform Code of Military Justice and on classified material and technologies.**

Copyright[753]

In addition to teaching and service, published research is a traditional mainstay of the academic profession. For faculty performance appraisals, all of the PME schools agree that while excellence in and dedication to teaching is the most important element of performance, they also expect research (publication) and service from the faculty. The PME institutions want to attract quality faculty, and faculty publication reflects positively on the PME institutions.[754]

Each PME school is using different copyright policies and practices. Some faculty reported not being able to publish materials in peer-reviewed journals or through academic presses because, as federal employees, they are barred by law from holding copyright of materials produced "as part of [their] official duties."[755] Federal law does not allow copyright protection for works of the United States Government, which are defined as "prepared by an officer or employee of the United States Government as part of that person's official duties."[756] Some PME institutions have issued policies on the circumstances in which professors at those schools may obtain copyright for the works they produce. Professors can copyright their work, if they do it at home, on their own time, with their own resources.[757] A separate but related consideration to the availability of copyright is whether outside sources can compensate government employees for works they produce. Ethics regulations promulgated by the Office of Government Ethics prohibit federal employees from receiving outside compensation for writing that relates to the employee's official duties; regulations define when this is operative.[758] The copyright policies at some PME institutions cross-reference or incorporate these regulations. The regulations, however, are specific only to determining outside compensation, not to determining copyright eligibility.

In other words, faculty members have to do something required for their jobs "off-duty." Faculty members may not be able to publish their work in the most reputable journals without copyright; most peer-reviewed journals require the author to turn over copyright in order to accept the material for publication. Some professors and institutions believe that if a professor is on sabbatical, and gains grant funds from outside sources, a copyright could be obtained. If the professor is on salary while on sabbatical, though, this interpretation could also be called into question. With sabbaticals at a premium, not everyone who needs to publish can be afforded sabbaticals, and sabbaticals can be garnered only once every five or six years, if at all. When asked about the impact of copyright laws on professors at CGSC, the school's official response was:

> The inability to hold copyright has in some instances acted as a constraint on faculty publication. For example, a faculty member who intends to publish with an academic press must do the writing on his or her own time. Alternatively, work done on duty time must be offered to a government press for first right of publication. Whether or not a government press chooses to publish the work in question, the inability to copyright is an impediment to outside publication since all academic and commercial publishers expect to copyright the works they publish. ...Needless to say, many civilian faculty chafe under this restriction and consider it a deterrent both to research and publication.[759]

USAWC professors also expressed concern over "the lack of copyright protection for government employees engaged in professional education and official research and publication." And, at National some professors stated that copyright issues are a problem affecting the school's ability to attract quality faculty.[760]

Having heard similar complaints, Representatives Platts and Skelton tried to address this issue through legislation in the 109th Congress in 2005 (See Appendix C.). They intended to encourage scholars to teach at PME institutions and to support their ability to publish. The legislation would have enabled PME faculty to assign copyright for their research to publishers as long as the faculty agreed to forego remuneration for work they did as official duty or with government resources.[761] The legislative effort was unsuccessful given the issue of wider application across the government.

This Subcommittee requested that the Congressional Research Service (CRS) examine the current copyright laws and suggest ways to approach these challenges. The following was submitted by CRS:

> A publisher of an academic paper would usually want to ensure that the paper is protected under the Copyright Act against unauthorized reproduction or distribution…a U.S. government work lacks such legal protection. Therefore, a publisher may not wish to publish an academic paper authored by faculty of military service academies or DOD professional schools due to the absence of legal remedies available to the publisher in the event of unauthorized reproduction or distribution of the published article. As a consequence, military service academies and DOD professional schools may encounter some difficulty recruiting or retaining faculty members who wish to establish or enhance their academic credentials through publication of their research papers in prestigious scholarly journals.[762]

To address any perceived disadvantages according to CRS, Congress could amend federal law in several ways: (1) To allow professors to hold an "unlimited" copyright (full copyright protection); (2) To allow professors to retain royalties or to assign royalties to a charitable foundation; or (3) To allow professors to have a limited copyright of their work for the sole purpose of transferring them to a scholarly journal that requires copyright as a condition of publication without professors gaining royalties or compensation.[763] However, any action on this must be recognized as a preferential regime for PME professors.

***Finding*: PME institutions have various copyright policies for their faculty members. Most faculty members are expected to publish scholarly works as a condition of employment and promotion. Allowing for at least limited copyright would contribute to PME institutions' ability to attract quality civilian faculty.**

***Recommendation*: The Subcommittee supports the reintroduction of H.R. 962 from the 109th Congress ("Platts-Skelton Amendment") or some appropriate variation of it in the 111th Congress.**

Faculty Development, Research and Sabbaticals, Divisions of Labor, and Guest Speakers

All of the PME institutions offer faculty development programs that are evaluated by the PAJE under OPMEP guidance.[764] Teachers and researchers are required to keep up with advances in their discipline and to network with other scholars in their field. This is particularly important to military and agency master's degree holders, but also to Ph.D.s who have not taught, have only taught undergraduates, or who have not taught in a seminar setting. To compound the challenge for instructors, some PME subject matter is unique, and most students are very accomplished professionals.

Each of the PME institutions has its own faculty development strategies and programs; there is no standardized approach. All have at least an orientation to the school and involve their faculty in beginning and/or end-of-year off-site sessions to explore lessons learned from the previous year, new curricula, and other issues. Most schools then allow individual departments to plan and manage their own introductory and continual faculty development. Schools also pair new instructors with experienced teachers for their first year if they have enough faculty.[765]

Because faculty development is left to the departments, instructors at a number of PME institutions look across departments and feel that they are not receiving as much benefit from faculty development efforts as some of their colleagues and peers.[766] While it is understandable that different departments and institutions have their own plans and processes for faculty development, this is an area that might benefit from comparison of best practices and common pitfalls both among departments and among schools.

Other faculty development is conducted on faculty members' own time. Professors and instructors undertake a wide variety of service and outreach activities for their institutions. All the schools direct that research not take priority over teaching but expect it to be done nonetheless. The reality is that most research and writing is generally done after duty hours. More significant research requires time away from the daily requirements of teaching such as during a sabbatical. Each of the PME institutions has different policies on sabbaticals as to years of teaching required before being considered and the length of a sabbatical. Again, while there is no need to standardize among schools, some faculty members voiced concern that application and selection criteria were not standardized and processes were not transparent within each institution. There were some allegations of favoritism.[767]

***Finding*: There are at least perceptions that the sabbatical processes at some PME institutions are not fair or transparent. This is particularly important because research and publishing are included in appraisals and recommendations for renewals, not to mention future employability.**

***Recommendation*: All institutions should review their policies and practices for sponsored research and ensure that faculty has access to documentation specifying criteria for application and selection procedures.**

In addition to research sabbaticals, both military and civilian faculty can compete for what are considered operational sabbaticals. For instance, instructors may take on research for the Secretary of Defense or the CJCS on a particular strategic or military problem. In addition, professors may deploy to study operational issues, or to maintain operational currency. In particular, permanent military professors, who would not be taking any future assignments in the operational force, were directed or decided to take "sabbaticals."[768] The Larson Report also suggested that permanent professors could fill fellowship positions at either their service's war college or National to enhance their professional development.[769]

Recent PAJE reports cover faculty development measures and recommendations for the intermediate and senior PME schools in some detail.[770] Institutions must program resources to support faculty service and research. These resources may come in the form of additional faculty to support the absence of those on sabbaticals and in the form of funds to sponsor travel and administrative costs.

In the education profession as in others, there is the dilemma of not having enough time for everything. PME faculty members are at various stages of their career, and as previously discussed, they vary from military operators and military academics to civilian agency representatives and civilian academics. The one thing that is common and constant at all PME institutions is that the first and most important task of a faculty member is teaching. But other factors come into play. Besides deployments and contributing to one's military service, there are "additional duties" expected of civilian and military faculty at PME institutions such as mandatory events that require faculty-to-faculty and faculty-to-student interactions that comprise an important part of the PME experience.[771]

Service is expected of all faculty members, but it may come in different forms depending on the person's rank, experience, and status. Military master's degree-level faculty are expected to focus on teaching excellence and mentoring students while also acting as uniformed role models at the lower-level schools. Military Ph.D. faculty members will spend more time on research and writing and on outreach and service. Civilian Ph.D.s concentrate on research, service, and outreach. Students sometimes play a role, too, in this constellation of activities. In the sciences, and perhaps the social sciences, students can participate in faculty research. In the humanities, however, that is much less likely. This is a factor in determining how much time and energy faculty members are able to expend on research. Students in the sciences essentially can act as research assistants.

For academics (Ph.D.s and terminal degree holders), whether military or civilian, teaching requires keeping current within their discipline. Master's level instructors will depend on senior academics for the newest discoveries, theories, and ideas. In order for academics to keep current, they must research, write, attend conferences, and network with their peers in the wider academic community. They must add to the knowledge in their field and create new knowledge. Research is not ancillary to the mission, but part of it. For recruiting and retaining civilian faculty, this emphasis and support is absolutely fundamental; it cannot be posed as an "either teaching or research" dilemma. Schools must have enough faculty members that all requirements, including research and writing, can be met. This takes institutional support – time and funding. Therefore, PME institutions must include this support requirement when managing their plans, personnel, and budgets.

Finally, PME institutions tend to entertain a large number of guest speakers to contribute to learning. Some are focused on illuminating particular lessons, but others are practitioners (senior officers or policy-makers). Some course directors told the Subcommittee that the budget for honoraria for speakers is severely limited.[772] Even the schools within the Capital region are not always able to bring in speakers they believe would enhance their mission. Speakers who "drop in" on schools engender the opposite problem. They are "free" but may not fit with the particular subject matter being covered; often the institution is not able to screen what the speaker will discuss. Many instructors and leaders of PME institutions observed that the number of these speakers should be limited to those who can really add to the student experience rather than be disruptive to the course of study. Unsolicited speakers simply add to the student and faculty contact hours reducing "contemplative" and research time. Some actually impart counterproductive messages. Students and faculty repeatedly cited examples of speakers suggesting the school year would be a great time to improve student golf scores.[773] Finally, the opposite issue also arose. Persons having special insight into course material or current events, might be on campus, but the students and faculty were not included on that visitor's agenda. A chance to capitalize on the person's presence was not realized.

The PAJE acknowledged that the Marine Command and Staff College's proximity to the Capital allowed for a robust guest speaker program and that this lent itself to increasing the students' interagency exposure. In fact, the PAJE report cited that guest speaker program as a "model for the other colleges."[774] The PAJE also commended Air War College for its speaker program, including the new Commandant's Lecture Series for providing "a wide variety" of speakers who discussed leadership issues from a "joint and interagency perspective."[775]

Marine Corps War College's speaker program was similarly commended.[776] On the other hand, the situation at National and ICAF was problematic. While the PAJE recognized that the individual schools' course lectures along with the Commandant's Lecture Series and NDU Distinguished Lecture Program all added value, the "sequencing and deconflicting of multiple guest lecture programs is problematic" and "the combined volume of speakers detracts from the overall programs and creates scheduling uncertainty."[777] The PAJE suggested that the schools and NDU establish a coordinated guest lecture policy that "limits lectures; disciplines the scheduling process; and allows for planning, scheduling and deconflicting guest lectures."[778]

Observation: A delicate balance must be maintained by course directors and school leaders as guest speakers' participation in PME should not be uncritically accepted because of availability, nor uncritically ruled out because of funding issues.

Support Staff

Leaders and faculty members at several of the intermediate- and senior-level institutions expressed concerns about being able to fit in requirements for research and outreach in addition to teaching and service. Some high-quality civilian faculty members were concerned that they would not have research assistants to help enable the work that would keep them current both in their subject matter and teaching.[779] Course directors, specifically, cited the lack of staff assistants to help with course administration, in addition to their faculty development and teaching load, not to mention requirements for them to do research and writing.[780] The Marine Corps War College stood out in this regard, being positively cited by the PAJE for adding an administrative program manager and other administrative and support personnel to reduce faculty workload in these ancillary areas.[781]

This is another relevant area for investing in human resources. If the smaller investment is made at the lower levels of assistance, higher paid faculty and leaders can concentrate on the tasks that only they can perform, and are required to perform for their appointment, retention, and promotion.

***Finding*: The mid-1990s saw, along with a total reduction in forces by almost half and an increase in PME students, a reduction of lower-ranking military and civilian administrative staff at many PME institutions as a cost-savings measure. Faculty and higher-ranking staff now perform administrative work that can detract from their focus on teaching, research, and service.**

***Recommendation*: Faculty requirements for administrative work (such as course direction), research and writing, and service and outreach should be matched with appropriate resources (time and money). Time can be gained not only by hiring the appropriate number of instructors but also by providing assistance for administrative work, if not also for research.**

Students

The Skelton Panel recognized the student body as another part of the "bedrock" foundation of the PME schools.[782] In previous portions of this report the students have been mentioned numerous times either as participants whom the Subcommittee interviewed or as the beneficiaries of this educational enterprise. Graduates will in turn make their contribution to their profession and the military as an institution. This section addresses U.S. military and civilian students. The former includes active-duty and reserve-component officers, and the latter includes DOD civilians, other U.S. Government or interagency civilians, and private industry students. PME schools also host a number of international military officers and foreign government civilian students.

Military Students

The Skelton Panel studied the PME student selection processes in detail. The Panel found that "there was a wide variance among the services' processes," and noted some issues concerning student quality. The Navy, for example, was noted for its process of identifying a pool of officers eligible for in-residence PME, relatively few of whom actually attended, rather than using a quality selection board. Overall, however, the Panel found that by and large, "the services select very capable officers for in-residence PME."[783] Nevertheless, some of the current quality concerns reflect the same issues that the Skelton Panel addressed more than 20 years ago.[784]

At the pre-commissioning level, students at the service academies are competitively selected from among the population and are chosen using a composite of qualifications ranging from academic potential to leadership. At the primary level for the Air Force, virtually all officers attend the Air and Space Basic Course (six weeks) for second lieutenants and Squadron Officers School (five weeks) for captains. Officers can complete the latter through distance education if their base commander does not select them, or if they do not have time to attend in-residence.[785] All Marine junior officers are required to complete Expeditionary Warfare School either in-residence or through distance learning.[786]

At the intermediate- and senior-levels, there are a number of variations as to how students are selected for the resident courses. The Army competitively selects officers for senior schools and fellowships, but in 2005 it implemented a policy of sending all majors to some form of in-residence intermediate PME, either for 10 months at Command and General Staff School (Leavenworth) or a shorter duration course at a satellite campus. In part, the Army's rationale was that as the security environment has become more complex, there is an increasing need for all officers to experience the benefit of some in-residence intermediate PME. Practically, however, the Army is only able to send about 75 percent of its majors to the course at Leavenworth. Combat arms majors generally attend the 10-month course, while their functional community counterparts study the "Common Core" at one of four satellite campuses. These students complete the remainder of the curriculum through distance education off-duty.

A number of Army students stated that they thought the lack of a selection board process for intermediate-level school resulted in the 10-month in-residence course at Leavenworth not being as rigorous or as challenging as what they perceived it used to be or what they thought it should be. Students pejoratively referred to the Army's commitment to universal intermediate PME as "no major left behind."[787] Faculty and current and former Army leaders at both the ILE and SLE levels noted concerns about whether combat arms officers were either being promoted so fast that they missed an opportunity to attend the course or were deployed so often that operational tempo did not allow for their attendance either at Leavenworth or other schools. Across the board, commentators expressed concern that this meant that some of the highest quality Army officers could reach senior rank without operational- or senior-level PME and, possibly, without JPME.[788] Senior leaders at Command

and General Staff College told the Subcommittee that the Army is rethinking the viability of universal intermediate PME.

Air Force and Marine Corps students are competitively selected either with, or separate from, their promotion boards for both intermediate- and senior-level PME schools. However, the Navy lacks a board-selection process for these schools. Navy promotion boards designate the top 50 percent of the lieutenant commanders (O-4s) and commanders (O-5s) as "school assignment eligible."[789] On occasion, officers falling below the 50 percent line may be selected for in-residence school. In his testimony before the Subcommittee, the President of the Naval War College highlighted a recent Navy policy that requires all O-5's taking command to have completed ILE (which includes JPME I) either in-residence or by distance education.[790] The Navy therefore, intends to encourage all officers who want to compete for command to complete some form of intermediate PME, whether through in-residence PME, at graduate school, or in distance learning courses.[791]

All of the schools reported that Air Force and Marine Corps students are top performers. The consensus among faculty and students at all of the ILE and SLE schools is that Navy students vary more widely in terms of quality and may either be far too junior (intermediate schools) or far too senior (senior schools) relative to their peers.[792] Faculty and students reported to the Subcommittee a number of times that Navy students were on the verge of retirement when they were sent to SLE.[793] Army students were seen as top performers at joint and other service schools; however, combat arms soldiers are being given fewer opportunities to attend in-residence PME, which has begun to affect this view.[794]

In order to have a standard and higher caliber of students across the services, some faculty members and experts believe the services should screen officers for SLE attendance, selecting only those officers who are academically talented. As a corollary, they believe that courses should be extremely rigorous and, like top-tier graduate or law schools, demand more from these officers during their academic assignments.[795] Other faculty and experts believe that officers who have proven themselves as outstanding operational leaders should be afforded the opportunity to attend senior schools in-residence. They assert that each student, whether academically inclined or not, should be brought to their individual best level of performance during their PME experience.[796]

This last issue centers on whether in-residence PME is desirable and feasible for all officers, or whether only those who are thought to have potential to reach the highest grades and positions should have this opportunity. The Army has decided to send all majors to ILE in-residence so that all O-4s have this baseline of education. The other services say they lack the capacity, or operational demands are too high, to do this. Even allowing for the recent initiative requiring intermediate PME in order to be competitive for command, Navy service culture still emphasizes the operational and technical demands of a naval career over PME.[797] The Skelton Panel took the position 20 years ago that this is not only a problem for the Navy, but that it also impacts the other intermediate and, in some instances, senior schools because of the varying quality and number of Navy students and faculty at these schools. All of these schools require at least some sea service representation in order to grant JPME credit.[798]

Finally, Subcommittee members were initially concerned that students were being pulled out of class early to meet operational requirements. While this did occur in some cases in the 2003-2004 timeframe, the services and schools have worked hard to keep these occurrences to a minimum. All of the schools have tried to frontload the most important requirements at the in-residence courses, so that if one or two students, on a case-by-case basis and approved at the highest levels, must be released, they can still gain the most important content and get credit for completing the course.[799]

Observation: Since 2003, very few students have been removed from PME courses early. PME leaders have argued strenuously against early release, and the schools have tried to minimize the impact of even the few early releases by ensuring that all JPME requirements are front-loaded in their curricula.

***Finding*: The Army's commitment to universal intermediate in-residence education since 2005 has had the unintended consequence of excluding many highly-competitive combat arms officers serving multiple tours in Iraq and Afghanistan from attending because of their limited windows of time to attend and because the Army has to push larger numbers of officers through school without a large increase in infrastructure and resources.**

***Finding*: The Navy remains alone among the services in not using a quality selection board process for in-residence PME students.**

***Recommendation*: The Navy should consider instituting a quality board process for student selection.**

Reserve Component

The Subcommittee heard that there are very few in-residence PME billets available to Army Reserve Component (RC) officers notwithstanding their significant contribution to current operations in Afghanistan and Iraq.[800] While reserve component officers should be afforded more opportunities to attend the shorter courses, RC officers cannot always afford to attend 10-month in residence courses because of their civilian employment, particularly if they have taken one or more long deployments.

Civilian Students

Interagency and DOD civilians do not have educational systems analogous to PME. The PME schools at the intermediate level, and more so at the senior level, offer select civilians a chance (usually just one) to attend courses with their military counterparts. The Subcommittee learned from faculty and students that the caliber of civilians attending officer PME institutions is uneven.[801]

Faculty and civilian students noted that agencies and services did not have a utilization plan for their graduates to capitalize on their PME experience. They often returned to the same position they left, or they found that they had to find a new position themselves after they completed PME because someone had either been hired into their former position or it had been deleted.[802] The Army and the Air Force have developed alternative approaches. The Army established a Graduate Placement Program in 2003 to ensure that its civilians were placed in "new positions requiring PME knowledge and skills" following completion of PME.[803] In 2009, 92 percent of graduates were placed in these new positions.[804] The Air Force requires civilian participants in PME to sign "mobility agreements," which require that they be willing to move upon completion of their PME. This, along with the Air Force's policy of centrally funding civilian PME billets (allowing a civilian's former office to hire a replacement), helps to ensure that civilian graduates are actively placed in new positions that utilize their new skills. The Air Force has also just begun a study into civilian follow- on assignments, and their Civilian Force Development Panel (made up of senior Air Force officers and members of the Senior Executive Service) will review that assignment information and consider improvements in finding suitable follow-on positions for civilian PME graduates.[805] In general, however, the services should have a clearer and more systematic approach to civilian attendance in PME and placement after PME graduation.

***Finding*: With some exceptions, government civilian PME graduates are not given follow- on assignments that recognize the value of their in-residence experience.**

***Recommendation*: The Department and the services should consider the Army and Air Force programs as models for enhancing civilian professional development through PME.**

DOD Civilians

In terms of quality, DOD civilian students were considered to be a very "mixed bag," with the best reputations being among OSD students but the worst reputations for quality being among the Defense Leadership and Management Program (DLAMP) students.[806] That program will close at the end of FY 2010, and a successor program, the Defense Senior Leadership Development Program (DSLDP), will take its place in order to improve student quality and contributions through competitive selection processes. The new DSLDP will "provide structured learning opportunities to enable the deliberate development of a diverse cadre of senior civilian leaders with…competencies needed to lead organizations…and achieve results in the joint, interagency, and multi-national environments."[807] Similarly, the FY 2010 NDAA established the Defense Civilian Leadership Program (DCLP) which was designed "to recruit individuals with academic merit, work experience, and demonstrated leadership skills" who would be eligible for rapid advancement and groomed for leadership positions within the Department. They would be given educational and training opportunities and placed in positions for which there is a definite need for their qualifications and abilities. Again, a competitive selection process is meant to ensure that only the most qualified participate in this program. While the emphasis of this program is placing the most qualified personnel in important and career-enhancing positions, the participants' access to educational opportunities could potentially raise the quality level of civilians that the Department sends to PME schools.[808] The impact of the DSLDP and DCLP cannot be analyzed yet.

Observation: Defense Leadership and Management Program (DLAMP) students attending PME, in general, did not have the same breadth of experience and were not of the same quality as their military counterparts. The old DLAMP program is set to close, and the new program is reportedly utilizing a quality board selection process. The Congress had such significant concerns about civilian personnel development that it included a requirement for the Department to develop the DCLP in FY 2010 NDAA.

Interagency Civilians

All of the intermediate and senior PME schools, except ICAF, National, and Marine C&SC, decried the lack of interagency students in their seminars. These students are highly valued for their alternative perspectives on national security and differing operational approaches and insights, particularly since all graduates will likely be operating in the joint, interagency, intergovernmental, and multinational (JIIM) environment. In fact, most PME leaders, faculty, and students desire that interagency participation be expanded beyond the "traditional national security" few of State, U.S. Agency for International Development (USAID), and the intelligence community. The leaders of the PME institutions understand the lack of a "personnel float" (or additional personnel who can be assigned to training) in these agencies and departments, but PME students largely believe greater numbers would improve their school experience and contribution to mission success later in the field and on staff.[809]

In this regard, the Subcommittee applauds the Army's approach to attracting interagency students to GCSS at Fort Leavenworth. Lieutenant General William B. Caldwell IV began implementation of a program through which the Army would offer a military CGSC graduate a fellowship to fill a desk at the State Department or USAID if the Foreign Service Officer (FSO) in that position were selected to attend CGSS. Further, the Army would provide priority for on-post housing for these FSOs so they would not have excessive out-of-pocket expenses for their move. This was all to be done on a non-reimbursable basis. Known as the "Casey Initiative" (after Army Chief of Staff General George Casey), this is seen as a win-win-win accomplishment. The Army officer benefits from a year-long interagency fellowship in which to put his or her education to work. The Foreign Service Officer completes PME in-residence with his or her military counterparts. And, the class has a larger component of interagency representation, which benefits all of the military students.[810] LTG Caldwell reported thinking about how to manage a similar program with students from non-governmental organizations in the future.[811] The Air Force sends officers to fellowships for interagency exposure but does not ask for interagency faculty or students in return.[812]

Two complaints emerged that should be addressed at the departmental level. Particularly at ILE, there have been instances when the Department of State has sent diplomatic security officers to PME. While they may be outstanding professionals and benefit from the curricula and association with their military classmates, their classmates do not gain perspectives on diplomacy and development that a Foreign Service officer would bring.[813] Coast Guard officers are also being counted as either interagency participants or sea service participants, depending on what the school needs most. This flexibility in counting students for OPMEP purposes does not reflect the contribution these sea service officers make; they are instead asked to represent the Department of Homeland Security.

Finding: **Service schools (other than the USMC, which has sufficient participation due to proximity to Washington, D.C.) see a need for more interagency faculty and students.**

Recommendation**: The services may want to consider the "Casey Initiative" being implemented at CGSC as a model to enroll more interagency students. Military officers would be sent post-PME to a one-year fellowship to a State Department, USAID, or other agency position. This, in effect, creates a *de facto* personnel float, enabling that department or agency to send a faculty member or student to the school in exchange for a year or more.**

International Students: Military and Civilian

Faculty and staff at most schools report that their International Fellows (IFs) bring a much needed perspective to the educational experience of U.S. officers and government civilians, but they, too, are a very mixed group. Those with the best English language skills contribute and gain the most.[814] While security assistance program contributions to allies and relationships with international partners should be a consideration, PME faculty indicated that there is a limit to the number of IFs who can be integrated into a seminar if teaching and learning for others were to be effective.[815] For that reason, adding more IFs does not necessarily improve the education for a JIIM environment. The Joint Staff also indicated that as more international students are added, the OPMEP-required representation of host service and non-host U.S. military students may be affected, ultimately affecting JPME accreditation.[816] Those who run the international officer programs noted that sometimes there is tension between the institutions' PME mission for U.S. students and the goals of the security cooperation mission under which the IFs are included.[817] These issues are being worked among the leaders of the schools, JCS/DJ-5, and the Military Education Coordinating Council.[818]

While all schools have a minimum Test of English as a Foreign Language (TOEFL) requirement to enroll foreign students, the schools do not all have the same requirement. Some school leaders and faculty thought this should be standardized.[819] In addition, International Fellows also miss a number of class sessions in order to participate in separate programs such as domestic travel but, except for the Naval War College where they have their own school, IFs earn the same PME certificates and master's degrees as their U.S. counterparts. At some schools, the IFs are subject to the same academic requirements and evaluations. At others, the PME leaders are sensitive to grading, evaluating, or "failing" a future minister of defense or chief of military from an allied or partner nation.

Finally, although most core courses are completely unclassified, security issues are raised in relation to international students. Some schools do not honor NATO officers' clearances even if they will be going to assignments where they will work with U.S. forces and utilize those same clearances.[820] In these situations, international students could clearly benefit from participation in those courses and activities that would help them in their next posting. U.S. students would also benefit from their presence and perspectives, just as they do during participation in the unclassified courses and events. Army War College faculty reported that they thought international students were not allowed to hold their clearances because of State Department policy.[821] Naval War College faculty said their institutional policy allowed them to receive waivers when this type of situation occurred in electives. They did not believe the State Department controlled this issue.[822] And at NDU, the IF management office policy is to treat all IFs the same – none will have access to classified material whether they have the appropriate clearances or not – so as not to separate the NATO students and other close partners from those who have not yet received that status. The Subcommittee heard complaints on this issue from international and U.S. students at the Army War College and at NDU.[823]

While adding more foreign military and government students should be considered, there is a limit to the numbers that make sense if the mission focus remains on U.S. military students' professional education, and if resources and infrastructure do not increase significantly. The implementation of the IF program does not have to be standardized necessarily among the services, but JCS/DJ-5 and DJ-7 should be included in coordination, and the school leaders and Military Education Coordinating Council should discuss where standardization might enhance both the educational opportunities of the U.S. and foreign students and the U.S. security cooperation function.

Industry Students

Because of ICAF's unique curricular focus on resourcing the national security and military strategies, the country's industrial base, and individual industries, ICAF includes students from the private-sector whose companies pay their tuition. However, as of 2009, only half the 20 seminars had private-sector students because the law limited NDU to a total of 10.[824] Staff, faculty, and students were unanimous in praising the contributions these students make to their military and civilian government peers' development and called for an increase in the number set by law to allow for one private sector student per seminar.[825]

Finding: **Although they have tried to mitigate the negative implications of too few industry students, ICAF argued that they could not provide the same quality of education on the industrial base and industries across the entire student body since only half the seminars included such private industry students. The FY 2010 NDAA included a measure to increase the number of private industry students allowed for ICAF by an additional 10.**

The theme that remains throughout this discussion on students is quality. If these schools are to produce future military leaders, attention must be paid to the types and mix of students that enter the classroom. These students should be upwardly mobile in their careers and should learn with fellow officers who share their intellectual curiosity with regard to operational art and strategy. The issue of quality does not stop with uniformed students but is also essential for civilian students and international and industry fellows. As the military increases its interagency interaction, it becomes necessary for military officers to learn from adept diplomats, foreign aid officers, and intelligence officers, among others. A student body that is comprehensively made up of quality military and civilian personnel becomes self-reinforcing, as the knowledge base and capabilities of all students increases due to their interaction with each other. To ignore the quality of students from any one military branch or sector of the civilian student body is to do all the others a disservice.

Material Resources

The Skelton Panel did not take a comprehensive look at how well the PME institutions were funded to accomplish their mission. The Panel did, however, inquire into the cost per student at each school. The Skelton Panel received from the Office of the Secretary of Defense raw data produced with different methodologies by service, and sometimes by school, which resulted in widely varying costs for roughly similar programs.[826]

For example, the Department submitted the fiscal year cost for a student at National War College as $9,387. The figure provided for a student at the Air War College was $121,348, the difference accounted for in large measure by the Air Force including student pay and moving costs associated with the permanent change of station. The Panel recommended that the Department "establish a uniform cost accounting system for the PME schools."[827]

The Subcommittee made its own effort to ascertain whether a uniform cost accounting system existed. The Department provided cost-per-student figures with standardized criteria. Comparative figures were no longer characterized by such enormous variations, but there were still a number of inexplicable differences.

The fiscal year 2009 cost per student for the five-week Air Force SOC was $94,711, almost as much as the $94,474 that the Marine Corps gave as the cost for the 10-month Marine Corps War College program.[828] There were as many as seven different variations allowed in computing the costs.[829] It is obvious that the Department had simply accepted and forwarded the services' numbers without providing and insisting on standardized procedures, or conducting its own analysis.

***Finding*: The Department does not have a uniform cost accounting method for the PME schools, and it has not provided figures that support useful comparisons.**

***Recommendation*: The Department should report to Congress on PME funding using a standardized accounting method for cost per student at each of the PME institutions as recommended by the Skelton Panel in 1989.**

The OPMEP has a Common Education Standard related to material resources. The standard mandates that "[e]ach institution must have a library or learning resource center, informational resources, financial resources, and physical resources that meet the needs of all users and support the mission and programs of the institution."[830]

Accordingly, PAJE reports routinely comment on libraries, information resources, and physical infrastructure in need of renovation, especially when the quality of the student experience is affected.[831] A recent report cited seminar rooms at one institution designed for 10-12 students that were described as "overcrowded" with 17 students per room, with no plans to renovate or replace the facility.[832] In another case, the Military Education Coordinating Council (MECC) mentioned the limited number of computer workstations available to students. The MECC also noted the potential impediment for accessing relevant and necessary academic information that may be imposed by not having an ".edu" domain and by having to negotiate ".mil" firewalls.[833] The MECC commented favorably on the Army's Lewis and Clark Center for the Command and General Staff College at Fort Leavenworth calling it "state of the art."[834] The Marine Corps is building to a similar standard in its strategic plan for Marine Corps University.[835] This said, the Subcommittee originally identified issues at Expeditionary Warfare School with funding for sufficient laptops for the student body there. Support for this requirement is now programmed for fiscal year 2011, and the school is addressing IT infrastructure issues to ensure that laptops can be utilized effectively.[836]

School leaders are often understandably reluctant to mention material resource challenges in a formal setting, not wanting to run afoul of decisions by higher headquarters for prioritization of funding. The Subcommittee, however, made similar observations to those of the MECC on its visits to the PME institutions. Physical infrastructure is limited at some institutions, particularly the older ones such as the NWC, National, and Marine Corps University. In addition, faculty at the joint schools, which rely on service sponsorship to fund their facilities, reported that they receive insufficient advocacy on resources issues.[837]

***Finding*: Adequate material support is necessary for mission effectiveness. While PME is not poorly resourced, some material challenges are apparent. Challenges include needs for reliable funding for the joint schools, renovated infrastructure for the older schools, and appropriate information and educational technology for many schools.**

***Recommendation*: The Department and the services should appropriately resource the mission-specific material needs of the PME institutions.**

Issues for Further Study

The Subcommittee conducted this study with a limited scope. In the process, the Subcommittee learned of several related issues outside of that scope that may warrant closer examination. The Subcommittee also encountered several outside commentators who have recently written on the topic of professional military education (PME) or proposed more extensive reform of the PME system, schools, or curricula. Without weighing their respective merits, the Subcommittee offers the following issues and sources of commentary for their additional and related views on PME.

Enlisted PME

All of the services have developed PME for enlisted personnel, however, enlisted PME is not as extensive as PME for officers. The Department has one joint in-residence PME (JPME) course for the most senior non-commissioned officers (NCOs) called Keystone taught at National Defense University (NDU). In his testimony before the Subcommittee, Dr. John Williams recommended that the services focus on "the increasing professionalization of the enlisted force and consider how enlisted educational opportunities can better meet evolving security challenges."[838] In addition, the Subcommittee visited Marine Corps University leaders and spoke to faculty members and students from the staff NCO course.[839] What is done there is impressive, but a comprehensive review of all of the service and joint enlisted PME programs is needed.

Non-Resident PME

As mentioned in this report, the services all have non-resident or "distance-learning" courses for their intermediate- and senior-level schools. Officers can satisfy the requirement for JPME I by completing the non-resident curricula for their service intermediate-level school. Some schools employ blended-learning options in which students mix short periods of in-residence instruction with their distance learning programs. While the senior schools all have non-resident courses, current legislation does not allow them to award JPME II credit for the distance-learning programs. However, technological advances allow students and instructors to interact with each other via audio and video links in real-time. They constitute an improvement on the correspondence courses used previously. The enhancement suggests that blended-learning programs could be structured to facilitate some measure of joint acculturation. Portions of the in-residence courses might be presented through distance learning or distributed learning (at more than one campus) to allow for students to spend more time on in-depth treatment of subject-matter best addressed in a seminar setting. One witness argued, however, that the officers enrolled in non-resident courses often have demanding operational duties and that this situation may not provide the opportunity for systematic thought and reflection that occurs through the in-resident PME programs.[840]

National Security Professional Development Program

The other departments and agencies of the federal government do not have professional education systems comparable to the PME framework. While a few civilians attend PME institutions, there is no educational vehicle dedicated to preparing government civilians to work across organizational boundaries in the same manner JPME fosters cooperation and teamwork among the military services. The National Security Professional Development (NSPD) program proposal aims to address the absence of systematic interagency acculturation through education, training, and professional experience. A tentative academic framework is being developed, and it would consist of a year of education and training in interagency coordination at the strategic level.[841] The proposal envisions that national security professionals would subsequently serve as senior advisors to department secretaries and agency heads and as liaisons between these entities and the National Security Council.[842]

Dr. Janet Breslin-Smith advocates reviving the 1946 concept of a National Security Education consortium between National War College, the Industrial College of the Armed Forces, the National Defense Intelligence College, and a new College of Diplomacy and Development.[843] While Dr. James Jay Carafano advocates clearly distinguishing PME from national security education for interagency leaders, he also believes Congress should build on the 2007 Presidential Order 13434 that established the NSPD program by providing a suitable governance structure and oversight. He compares such an effort to the Goldwater-Nichols Act and military jointness.[844]

DOD Civilian Education

Civilians in the Department of Defense and its component departments face challenges similar to those of other government civilians. Although certain services offer brief management courses to their senior civilians in partnership with civilian universities and at the Federal Executive Institute, DOD civilian education lacks the equivalent of the primary- and intermediate-level PME institutional framework.

Changing Retirement and Promotion Policies

As addressed in this report, more flexibility is needed with respect to career paths and retirement and promotion incentives due to longer career possibilities. Changing policies will also impact retention. The Department, CJCS, and the services should study the likely impact of these changes. Longer careers combined with the need for personnel with low density and high demand skill sets including foreign area officers, civil affairs officers, language specialists, and strategists could create more time and opportunity for training, education, and developmental assignments. More personnel might be sent for foreign language study, doctoral programs at civilian schools, or PME faculty duty.

A Word about Ethics

Teaching Ethics

Arguably, no topic is more important and no topic more difficult to teach than ethics. Ethics and ethical behavior are fundamental to what our nation expects of its officers. Indeed, the three service academies emphasize ethics from the beginning of a cadet's or midshipman's experience and each makes moral and character development a central part of the academy mission. The Air Force Academy vision, illustrative of those at the other academies, is "to be the Air Force's premier institution for developing leaders of character."

Despite the common emphasis on honor and character among the academies, each academy honor code is unique to that institution. Notably, the Naval Academy code is the only one which does not consider tolerance of an honor code violation an honor violation itself. Both the U.S. Military Academy and the Air Force Academy codes state that cadets will not tolerate those who violate the code; those who do tolerate are also guilty of an honor violation. Instead, the Naval Academy Honor Concept makes a positive statement, that "Midshipmen are persons of integrity. They stand for that which is right." If a midshipman is found guilty of tolerating another's honor violation it constitutes a conduct offense under "failure to follow written orders."[845] This may seem to be a minor point, but in ethics debates it could have significant implications. One or another approach is not wrong or right, simply different.

Reflecting the distinct honor codes and moral development curricula at the three service academies, the services take differing approaches to developing ethical judgment in our military officers even at the most basic level. What is clear is that the military services believe that early training and continuous reinforcement of ethical leadership remains important. The difficult choices facing military leaders only become more difficult as they rise in rank and face more complex operational and strategic challenges.

While it is important that officers demonstrate ethical behavior in their personal conduct, the critical need is that officers understand the importance of setting ethical standards in the units they lead. For leaders, ethics involves enforcing standards of conduct at all times, especially under stressful conditions. An enduring area of difficulty arises when people violate established standards in their interactions with members of the opposite sex. Fraternization, sexual harassment, sexual assault, or other abuses of power are no more tolerable in a deployed setting than in garrison, yet persist with troubling regularity.

Since the integration of women into mainstream military units began in the 1970's, each of the services have experienced challenges and scandals involving the academies, training bases, and operational units. Reports of ignored incidents of sexual harassment, or far worse, sexual assault, among deployed units can be indicators of a failure in the chain of command at some level to set the proper example and creates the perception of condoning improper or criminal behavior. The proper enforcement of high standards of conduct is even more important when deployed because of the mutual trust necessary to operate in what can be a pressure-filled and dangerous environment. When operational pressures are intense, commanders can feel pressure to relax standards or rationalize handling serious incidences internally rather than through proper procedures.

The persistence of these reports, plus other emerging reports of misconduct from the recent war experiences involving prisoner abuse, contracting fraud, and other incidents, show that ethics concerns are relevant and real. The Subcommittee firmly believes that the vast majority of American service members conduct themselves honorably. Yet, the disproportionate impact that improper behavior has on the overall mission bears renewed emphasis at all levels.

Ethics Education Requires More Attention

Because the Subcommittee did not delve deeply into curriculum topics except as exemplars of how dynamic the PME curricula are in responding to a changing security environment, it did not examine the subject of the teaching of ethics to military officers in detail. There have been several reviews of military ethical lapses that have recommended in-depth studies of how PME institutions teach ethics. The 2004 Cheney Report on PME recommended that because of ethical issues at the academies there should be a comprehensive review of how ethics is taught at all the PME institutions.[846] The 2006 Schlesinger Report on Abu Ghraib also recommended a review of how ethics is taught in PME courses.

The current *Officer PME Policy* (OPMEP) includes a learning objective for senior-level school under "Joint Operational Leadership" to "comprehend the ethical dimension of operational leadership and the challenges it may present."[847] There is a similar objective for intermediate schools to "comprehend the ethical dimension of operational leadership and the challenges it may present." Every PME school with a joint component must address these objectives in its core courses in order to maintain joint accreditation. However, the JCS/DJ-7 maintains that teaching ethics is primarily a service and training function.[848]

Nonetheless, the OPMEP recognizes that ethics is part of the PME continuum with each level building on another, much as the Cheney Panel recommended.[849] At the pre-commissioning level, the OPMEP focuses on developing a foundation in "....leadership, management, ethics, and other subjects necessary to prepare them to serve as commissioned officers."[850] At the flag officer and senior levels, the OPMEP objectives for JPME "go to the skills necessary to build and sustain ethical organizations and to further evaluate the ethical ramifications of specific historical and contemporary national security decisions."[851]

The senior joint and service schools told the Subcommittee how their curricula focus on ethics.[852] Many of the PME institutions treat ethics by making it part of their elective offerings. Others include only one or two lessons in the core curriculum. Still others seek to weave the discussion and consideration of ethics throughout the core curriculum. While the last of these is the most difficult as it calls on all instructors to be well-versed, it is probably the most effective means of addressing this critical area.[853]

Cheney and Taylor argued that PME should and does reinforce the military ethos. They proposed that the treatment of ethics must go beyond what the lawyers, inspectors general, auditors, special counsels, and investigators have to know and cited numerous examples of lapses among the officer corps outside these arenas. They reported, "[e]ven a number of senior active-duty military leaders believe the intermediate and senior levels of the PME system have failed to emphasize an officer's ethical and moral development in an increasingly complex world adequately.[854] The Panel recommended the establishment of guidelines for a comprehensive, sequential, and progressive program because "officers need to refine how they think about important ethical issues, wrestling along with their fellow officers with the moral dilemmas presented by real-world case studies. ...[C]ase studies should be used, and it should be understood that some 'dirty linen' may be aired in the process."[855]

Finally, the Cheney Report addressed senior-level PME, when "the treatment of ethics should include rigorous examination of the ethical content of policies and strategies, not just of the values and behaviors of senior officers. Perhaps the most fundamental lesson to be instilled in officers is that nothing has greater impact on the ethical behavior of subordinates than the command climate that is created by senior leaders."[856]

NDU has established one of the newest ethics centers among the PME institutions.[857] General Richard B. Myers, former Chairman of the Joint Chiefs of Staff (CJCS), supports the work of the NDU Institute for National Security Ethics and Leadership as the Colin L. Powell Chair for National Security, Leadership, Character, and Ethics.[858] He agrees that the Cheney Report recommendation for a comprehensive review of ethics education in PME should be undertaken. One of the outcomes, Myers thinks, should be clear recommendations for improvement including legislation that would put the responsibility directly on the CJCS to report back to Congress with a plan to implement a comprehensive JPME ethics education program. The next step for Myers would be that the CJCS task his joint staff and NDU to implement the JPME ethics program, with responsibility for the service PME programs in ethics given to the Service Chiefs. Myers proposes that these programs should encompass the pre-commissioning and company-grade courses as well as the more senior levels. He adds that while the services and schools can tailor their programs, there should be a common curriculum that covers the fundamental issues so that the courses will not constitute just a "hodgepodge" of ideas and courses.

The Military Education Coordination Council (MECC), according to Myers, could coordinate the curricula and make revisions through an annual ethics conference. It is not enough to do what the PME schools do, asserted Myers, which is generally to have ethics covered in their "leadership module" and one elective taught by their "expert." Each school will still need an expert on ethics and electives and special classes, but ethics should also be blended into the entire core curriculum.

While Myers acknowledges that progress has been made over the years, more work is needed in this area. Admiral Michael G. Mullen, the current CJCS, reinforced this point in an interview in a recent edition of *Joint Forces Quarterly* dedicated to the subject of military ethics. Mullen states clearly, "I would agree that we do need more of a focus on military ethics and civil-military relations in our schoolhouses."[859]

While initial and reinforcing ethics education and training are necessarily a military service responsibility, today's complex operational environment would seem to call for even more targeted training that addresses the challenges faced by joint and deployed units today. The stresses have not diminished, and the complexity of situations that officers face has grown ever more daunting. Ethics education and training should keep pace with the demands placed upon our military leaders, as General Myers, Admiral Mullen, and others suggest.

Outside Commentary

Lieutenant General David Barno, USA (Ret.)

- "Senior Leader Development – Time for a 'College for Generals,'" draft manuscript, 25 February 2009.

John Collins

- "The Case for a Grand Strategy College," an unpublished monograph, 2009.

Captain Mark R. Hagerott, USN

- "Rebalancing the Naval Officer Corps," *Keeping the Edge: Revitalizing America's Military Officer Corps*, Center for a New American Security, February 2010.
- "The Heart of an Officer: Joint Interagency, and International Operations and Navy Career Development," *Naval War College Review*, Spring 2009.

Frank G. Hoffman

- "Embracing a Full Spectrum Profession," *Keeping the Edge: Revitalizing America's Military Officer Corps*, Center for a New American Security, February 2010.

Dr. Maren Leed and David Sokolow

- The Ingenuity Gap: Officer Management for the 21st Century, A Report of the Center for Strategic and International Studies International Security Program, January 2010.

General James N. Mattis, USMC

- "Developing and Retaining the Officers We Need for the 21st Century," remarks delivered to the Center for a New American Security, 18 February 2010.

Dr. John A. Nagl and Brian M. Burton

- "Revitalizing America's Officer Corps," *Keeping the Edge: Revitalizing America's Military Officer Corps*, Center for a New American Security, February 2010.

Major general Robert H. Scales, USA (Ret.)

- "Too Busy to Learn," *Proceedings*, February 2010.
- "Return of the Jedi," *Armed Forces Journal*, October 2009.

Dr. Don M. Snider

- "Developing a Corps of Professionals," Keeping the Edge: Revitalizing America's Military Officer Corps, Center for a New American Security, February 2010.

Admiral James G. Stavridis

- "The Heart of an Officer: Joint Interagency, and International Operations and Navy Career Development," *Naval War College Review*, Spring 2009.

Colonel Roderick C. Zastrow, USAF

- "Strategic Leader Development from an Air Force Perspective," *Keeping the Edge: Revitalizing America's Military Officer Corps*, Center for a New American Security, February 2010.

CONCLUSION

After a year of study and careful consideration of the United States' officer, in-residence professional military education (PME) system, the Subcommittee members are heartened by what we learned.

While we undertook this study convinced of the value of our professional military education system, we began with a fundamental question. What does PME contribute to officer development? The answer to that question goes to the heart of the American military tradition, and what it means to serve as a commissioned officer in the United States Armed Forces. From our country's birth, the United States has valued selfless, ethical officers who are adept at leading diverse groups in the execution of complex, dangerous missions. True to these beginnings, our professional military education system develops military officers along three axes: character, or ethical and moral leadership; acculturation, or learning from one's peers; and intellectual development, critical thinking, and mental agility. While we found that our PME system addresses all three of these important areas, improvements are needed in each.

We touch on ethics in the "Issues for Further Study" section of this report, and recommend a comprehensive review and renewed emphasis on ethics in PME. Acculturation has become even more important with the ascendancy of joint, interagency, and intergovernmental operations, making the opportunity to learn with students from State, Justice, and Homeland Security, among other agencies, even more important than before. Though ethical education and opportunities for acculturation need improvement, we are encouraged by the Department's plans and progress in these two areas.

As a result of our study, we also reassert the value of in-residence officer PME. The Subcommittee finds that the reinforcing nature of in-residence PME has value in inculcating officer and leader values in student officers. In addition, we find the opportunity to learn directly from others with varied backgrounds invaluable. There is value in getting away from operational responsibilities to bond with fellow officers, both from one's own service and from other services, and with civilian and international students. Intermittent breaks for in-residence education provide contemplative time to put operational lessons into context. These experiences are especially meaningful at the war college level, but are also valuable for more junior officers.

We find the greatest need for improvement in the third fundamental PME mission area, intellectual development. This report explores various means by which critical thinking skills can be improved in the officer corps as a whole and how more strategists may be developed. The Subcommittee finds that both developing critical thinking skills among all officers and cultivating a number of skilled strategists remain important objectives of our PME system.

Military officers have always been called upon to develop creative solutions to complex problems. The Subcommittee heard considerable evidence of the need for officers with strong critical thinking skills from the most junior to the most senior levels. If anything, officers are confronted with these challenges more often and at more junior ranks than ever before and, at times, with strategic implications riding on their decisions and their actions. We are concerned by the lack of a coherent, comprehensive, and effective program to improve critical thinking skills among all officers continually throughout their careers. We also found that there is a pronounced need for officers with strategic vision. In our view, the most gifted strategists are pragmatic innovators with whole-of-government and global perspectives. We found no uniform or coordinated plan for elevating well-educated strategists to critical decision-making positions at the most senior levels of command.

In addition to a lack of a comprehensive plan to improve critical thinking among all officers, we are also troubled by the gradual loosening of the link between joint education and joint assignments. What seems to be fraying in the relentless demands of wartime operations is a foundational link between joint education and preparation for joint duty assignments. That tie needs to be reexamined and reinforced. We have made several findings and recommendations about this issue, and look forward to working with the Department to address these concerns.

Finally, we examined various issues concerning the important human dimension of our PME system and touched on material resource issues as well. Although both affect the quality of the Nation's military institutions, we believe faculty issues deserve the most urgent attention. While we continue to benefit from outstanding faculty, students, and facilities, the Department and the Committee should address these challenges in order to strengthen the PME system. As the committee with oversight responsibility, the House Armed Services Committee will honor the legacy of the Skelton Panel by leading this effort in Congress.

In the last two centuries, the United States has developed an impressive group of strong institutions to ground and educate military officers in the finest traditions of our country. That system endures. Beyond maintaining the strength of the essential and enduring aspects of PME, and beyond rededicating the system to support the tie between joint education and joint duty assignments, we emphasize once more what is new. Professional military officers must not only have the highest integrity and be willing and able to lead others in facing danger and adversity, they must be agile of mind. They must be broadly educated to face a complex and uncertain environment beyond the military and beyond our shores. And, they must be able to communicate clearly not only with those they lead but with the citizens they serve, the partners with whom they work, and even with the adversaries whom they face.

While we have made suggestions for improvement, we were inspired by the commitment of those engaged in this important enterprise. We were especially impressed with the students and faculty we met this year. Our country's system for educating its officers remains sound. The Committee will continue to work with the Department of Defense to maintain that strength and improve it where we can.

Appendix A : Glossary of Acronyms

AAUP	American Association of University Professors
ACSC	Air Command and Staff College
AETC	Air Education and Training Command
AFB	Air Force Base
AFSC	Armed Forces Staff College
AMSP	Advanced Military Studies Program
ANSCOL	Army-Navy Staff College
AOWC	Advanced Operations Warfighting Course
ASBC	Air and Space Basic Course
AU	Air University
C&SC	Marine Corps Command and Staff College
CAC	Combined Arms Center
CAO	Chief Academic Officer
CCJO	Capstone Concept for Joint Operations
CCRB	Course Content Review Board

CENTCOM	Central Command
CGSC	Command and General Staff College
CGSS	Command and General Staff School
CIWAG	Center on Irregular Warfare and Armed Groups
CJCS	Chairman of the Joint Chiefs of Staff
CNA	Center for Naval Analyses
CNAS	Center for a New American Security
CNCS	College of Naval Command and Staff
CNW	College of Naval Warfare
COCOMs	Combatant Commands or Combatant Commanders
CRB	Curriculum Review Board
CRS	Congressional Research Service
CRWG	Curriculum Review Working Group
CSIS	Center for Strategic and International Studies
DASD	Deputy Assistant Secretary of Defense
DCLP	Defense Civilian Leadership Program
DDJS-ME	Deputy Director of the Joint Staff for Military Education
DG	Distinguished Graduate
DHS	Department of Homeland Security
DJS	Director of the Joint Staff
DLAMP	Defense Leadership and Management Program
DOD	Department of Defense
DSLDP	Defense Senior Leadership and Development Program
DT	Development Team
eJMAP	electronic Joint Manpower and Personnel System
EPRB	Educational Program Review Board
EWS	Expeditionary Warfare School
FA 59	Functional Area 59, Strategist
GDF	Guidance for the Development of the Force
GS	General Schedule
HASC	House Armed Services Committee
HQ	Headquarters
HRD	Human Resources Directorate
ICAF	Industrial College of the Armed Forces
IF	International Fellow
ILE	Intermediate Level Education
IT	Information Technology
IW	Irregular Warfare
JAWS	Joint Advanced Warfighting School
JCS	Joint Chiefs of Staff
JCS/DJ-7	Operational Plans and Joint Force Development Directorate of the Joint Chiefs of Staff
JCWS	Joint and Combined Warfighting School
JDAL	Joint Duty Assignment List
JDAMIS	Joint Duty Assignment Management Information System
JFCOM	United States Joint Forces Command

JFSC	Joint Forces Staff College
JIIM	Joint, Interagency, Intergovernmental, and Multinational
JOE	Joint Operating Environment
JOM	Joint Officer Management
JOMS	Joint Officer Management System
JOPES	Joint Operational Planning and Execution System
JPME	Joint Professional Military Education
JQO	Joint Qualified Officer
JQS	Joint Qualification System
JSO	Joint Specialty Officer
MAGTF	Marine Air Ground Task Force
MCPP	Marine Corps Planning Process
MCU	Marine Corps University
MDMP	Military Decision Making Process
MECC	Military Education Coordination Council
MMAS	Masters of Military Arts and Science
MOU	Memorandum of Understanding
National	National War College
NATO	North Atlantic Treaty Organization
NCO	Non-Commissioned Officer
NDAA	National Defense Authorization Act
NDU	National Defense University
NDU-P	National Defense University President
NETC	Naval Education and Training Command
NMS	National Military Strategy
NORAD	North American Aerospace Defense Command
NORTHCOM	Northern Command
NSS	National Security Strategy
NSU	National Security University
NWC	Naval War College
O&I	Oversight and Investigations
OPLANS	Operational Plans
OPMEP	Officer's PME Policy, CJCS Instruction 1800.01D
OSD	Office of the Secretary of Defense
PAJE	Program for the Assessment of Joint Education
PJE	Program for Joint Education
PME	Professional Military Education
PMP	Permanent Military Professor
PP	Permanent Professor
PRT	Provincial Reconstruction Team
QDR	Quadrennial Defense Review Report
RC	Reserve Component
SAASS	School of Advanced Air and Space Studies
SAE	Special Area of Emphasis
SAW	School of Advanced Warfighting
SLE	Senior Level Education

SOC	Squadron Officer's College
SOCOM	United States Special Operations Command
SOS	Squadron Officers School
SOUTHCOM	United States Southern Command
SSS	Senior Service School
STRATCOM	United States Strategic Command
TRADOC	Training and Doctrine Command
UCMJ	Uniform Code of Military Justice
USA	United States Army
USAF	United States Air Force
USAFA	United States Air Force Academy
USAID	United States Agency for International Development
USAWC	United States Army War College
USD	Under Secretary of Defense
USD (P&R)	Under Secretary of Defense for Personnel and Readiness
USERRA	United Services Employment and Reemployment Rights Act of 1994
USMA	United States Military Academy
USMC	United States Marine Corps
USN	United States Navy
USNA	United States Naval Academy
WMD	Weapons of Mass Destruction

APPENDIX B: ORGANIZATIONAL CHARTS

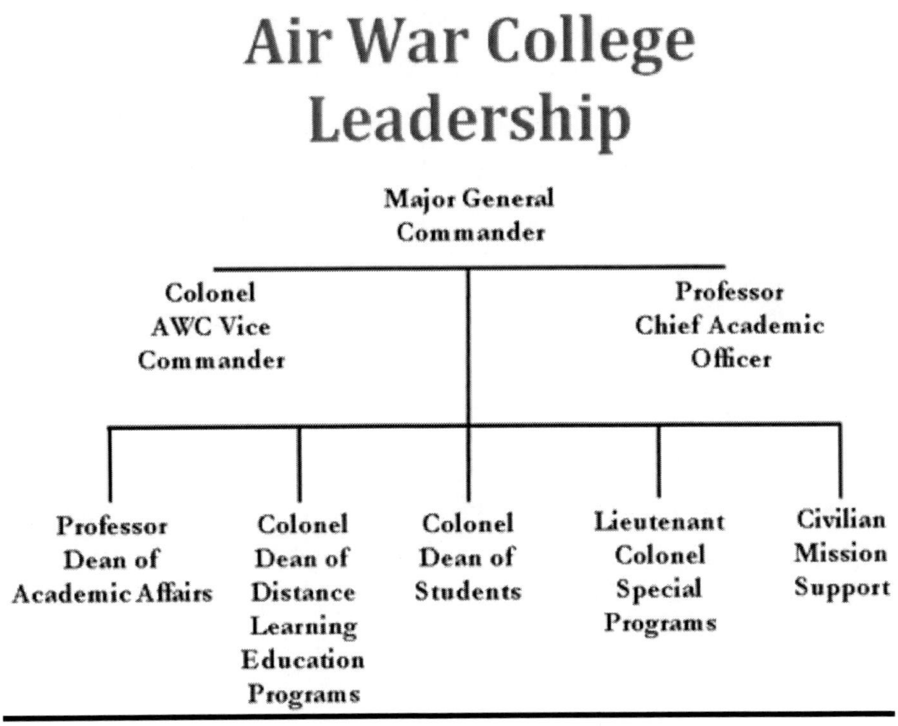

ACSC Leadership

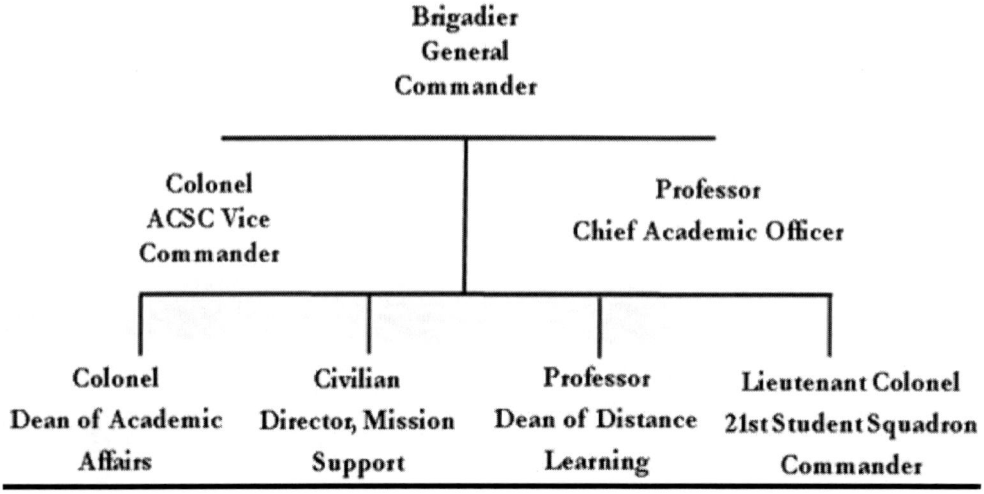

Squadron Officer College Leadership

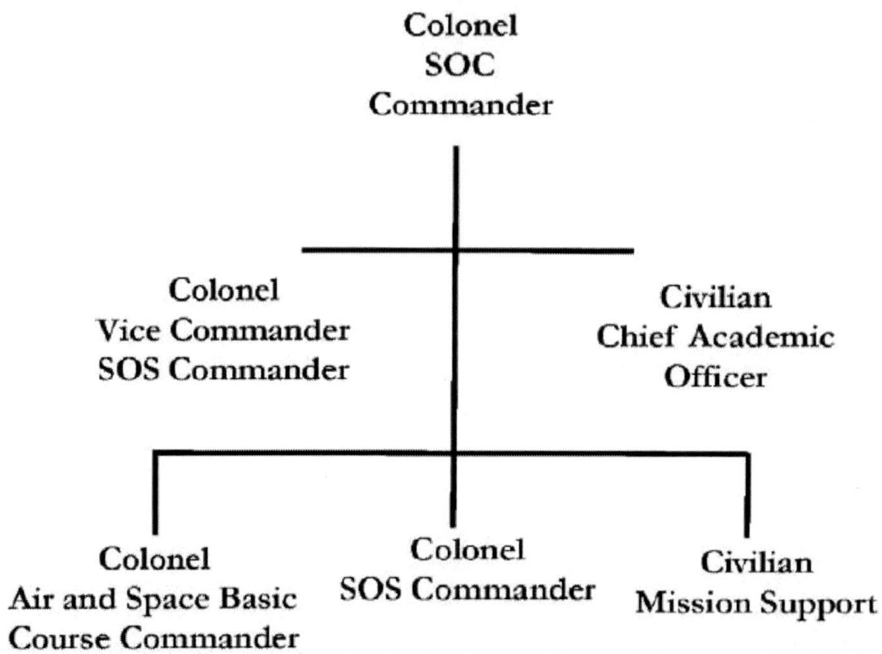

APPENDIX C: AMENDMENT ON COPYRIGHT

109TH CONGRESS
1ST SESSION

H. R. 962

To amend title 10, United States Code, to allow faculty members at Department of Defense service academies and schools of professional military education to secure copyrights for certain scholarly works that they produce as part of their official duties in order to submit such works for publication, and for other purposes.

IN THE HOUSE OF REPRESENTATIVES

FEBRUARY 17, 2005

Mr. PLATTS (for himself and Mr. SKELTON) introduced the following bill; which was referred to the Committee on the Judiciary, and in addition to the Committees on Transportation and Infrastructure and Armed Services, for a period to be subsequently determined by the Speaker, in each case for consideration of such provisions as fall within the jurisdiction of the committee concerned

---

A BILL

To amend title 10, United States Code, to allow faculty members at Department of Defense service academies and schools of professional military education to secure copyrights for certain scholarly works that they produce as part of their official duties in order to submit such works for publication, and for other purposes.

Be it enacted by the Senate and House of Representatives of the United States of America in Congress assembled,

SECTION 1. LIMITED AUTHORITY FOR DEPARTMENT OF DEFENSE PERSONNEL WHO ARE FACULTY MEMBERS AT DEPARTMENT OF DEFENSE SCHOOLS OR OTHER ACADEMIES TO SECURE COPYRIGHTS FOR CERTAIN SCHOLARLY WORKS.

(a) AUTHORITY.—(1) Chapter 53 of title 10, United States Code, is amended by inserting after section 1033 the following new section:

"§ 1033a. Faculty of service academies and Department of Defense professional schools: limited authority to secure copyrights for certain works

"(a) AUTHORITY.—Subject to regulations prescribed under subsection (f), a person who is a member of the Army, Navy, Air Force, or Marine Corps, or a civilian employee of the Department of Defense, and is a faculty member of an institution described in subsection (e) may, notwithstanding section 105 or 201(b) of title 17, secure copyright protection under title 17 for a qualifying work, but only for the purposes of submitting such work for publication in a scholarly journal, publication, or other edited work for which such a copyright is a requirement for consideration for publication or otherwise as may be prescribed under regulations under this section.

"(b) QUALIFYING WORKS.—A work is a qualifying work for purposes of this section if the work—

"(1) is prepared as part of a person's official duties; and

"(2) meets such criteria as the Secretary of Defense may prescribe by regulation as a scholarly work for which copyright protection as provided in subsection (a) is warranted.

"(c) TRANSFER OF COPYRIGHT.—Upon acceptance for publication of a work for which copyright protection exists by reason of subsection (a), the person holding the copyright shall transfer the copyright to the owner or publisher of the medium in which the work will be published.

"(d) ROYALTIES, ETC.—No royalties or other compensation may be accepted by a person described in subsection (a) by reason of copyright protection that exists by reason of subsection (a).

"(e) COVERED INSTITUTIONS.—The institutions referred to in subsection (a) are the following:

"(1) The United States Military Academy, United States Naval Academy, and United States Air Force Academy.

"(2) The National Defense University.

"(3) Any war college of the armed forces.

"(4) Any graduate-level college or university of the Department of Defense.

"(5) The Coast Guard Academy.

"(6) The United States Merchant Marine Academy.

"(f) REGULATIONS.—The Secretary of Defense shall prescribe regulations for the purposes of this section. Such regulations shall include provisions specifying the types of works for which copyright protection may be secured by a person described in subsection (a).".

(2) The table of sections at the beginning of such chapter is amended by inserting after the item relating to section 1033 the following new item:

"1033a. Faculty of service academies and Department of Defense professional schools: limited authority to secure copyrights for certain works.".

(b) EFFECTIVE DATE.—Section 1033a of title 10, United States Code, as added by subsection (a), shall apply only with respect to works that, as determined under regulations prescribed under that section, are produced after the date of the enactment of this Act.

(c) DEADLINE FOR REGULATIONS.—The Secretary of Defense shall prescribe regulations under subsection (f) of section 1033a of title 10, United States Code, as added by subsection (a), not later than 180 days after the date of the enactment of this Act.

Appendix D: Hearings, Meetings, Briefings, Travel

O&I Hearings

Another Crossroads? Professional Military Education Twenty Years after the Goldwater-Nichols Act and the Skelton Panel, 20 May 2009.

Witnesses:

Dr. Janet Breslin-Smith
Retired Professor and Former Department Head
The National War College
Dr. Alexander Cochran
Historical Advisor to the Chief of Staff of the Army
U.S. Department of the Army
Dr. James Carafano
Assistant Director, Kathryn and Shelby Cullom Davis Institute for International Studies
The Heritage Foundation
Dr. Richard Kohn
Professor of History, and Peace, War, and Defense
The University of North Carolina at Chapel Hill

Thinkers and Practitioners: Do Senior Professional Military Education Schools Produce Strategists? 4 June 2009.

Witnesses:
Rear Admiral Garry E. Hall, USN
Commandant
The Industrial College of the Armed Forces
Major General Robert P. Steel, USAF
Commandant
The National War College
Rear Admiral James P. Wisecup, USN
President
The Naval War College
Major General Robert M. Williams, USA
Commandant
The Army War College
Major General Maurice "Maury" Forsyth, USAF
Commander of the Spaatz Center and Commandant
The Air War College
Colonel Michael Belcher, USMC
Director
The Marine Corps War College

Thinking from the Tactical to the Operational Level: JPME I and II at the Services' and Joint Command & Staff Colleges, 25 June 2009.

Witnesses:
Brigadier General Edward C. Cardon, USA
Deputy Commandant
Army Command and General Staff College
Brigadier General Katherine P. Kasun, USA
Commandant
Joint Forces Staff College
Brigadier General Jimmie C. Jackson, USAF
Commandant
Air Command and Staff College
Rear Admiral James P. Wisecup, USN
President
Naval War College
Colonel Raymond C. Damm, USMC
Director
U.S. Marine Corps Command and Staff College

Beyond Service Core Competency: Are Our Junior Officers Prepared for Today's Security Environment? 15 July 2009.

Witnesses:
Brigadier General Dana H. Born
Dean of the Faculty
U.S. Air Force Academy
Brigadier General Patrick Finnegan
Dean of the Academic Board
U.S. Military Academy
Captain Matthew L. Klunder
Commandant of Midshipmen
U.S. Naval Academy
Colonel Stephen M. Tanous
Commandant
U.S. Air Force Squadron Officer College
Colonel Brian D. Beaudreault
Director
U.S. Marine Corps Expeditionary Warfare School

Investing in Our Military Leaders: The Role of Professional Military Education in Officer Development, 28 July 2009.

Witnesses:
Lieutenant General John M. Paxton, Jr., USMC
Director of Operations/Acting Director of the Joint Staff

Joint Staff
Lieutenant General William B. Caldwell, IV, USA
Commanding General
Combined Arms Center
Deputy Commanding General
Training and Doctrine Command Fort Leavenworth
U.S. Army
Mr. Dan R. Sitterly
Director of Force Development
Deputy Chief of Staff
Manpower and Personnel
U.S. Air Force
Mr. Scott W. Lutterloh
Director
Training and Education Division
U.S. Navy
Brigadier General (MajGen select) Melvin G. Spiese, USMC
Commanding General
Training and Education Command
U.S. Marine Corps

Charting the Course for Effective Professional Military Education, 10 September 2009.

Witnesses:
Lieutenant General David W. Barno, USA (Ret.)
Director
Near East South Asia Center for Strategic Studies
Dr. John Allen Williams, Ph.D.
Professor of Political Science
Loyola University Chicago
President
Inter-University Seminar on Armed Forces and Society
Dr. Williamson Murray, Ph.D.
Professor Emeritus
The Ohio State University
Senior Fellow
Institute for Defense Analyses

Other Hearings

The full committee met to receive testimony on the priorities of the Department of Defense in the new Administration, 27 January 2009.

Witness:
The Honorable Robert M. Gates
Secretary of Defense

The full committee: Security developments in the areas of responsibility of the U.S. Southern Command, Northern Command, Africa Command, and Joint Forces Command, 18 March 2009.

Witnesses:
Admiral James G. Stavridis, USN
Commander
U. S. Southern Command
General Victor Eugene "Gene" Renuart, Jr., USAF
Commander
U. S. Northern Command and Commander,
North American Aerospace Defense Command
General James N. Mattis, USMC
Commander
U.S. Joint Forces Command and Supreme-
Allied Commander Transformation for NATO
General William E. "Kip" Ward, USA
Commander
U.S. Africa Command

The full committee: Security developments in the areas of responsibility of the U.S. Pacific Command, U.S. European Command, and U.S. Forces Korea, 24 March 2009.

Witnesses:
Admiral Timothy J. Keating, USN
Commander, U.S. Pacific Command
General Bantz J. Craddock, USA
Commander, U.S. European Command
NATO Supreme Allied Commander Europe
General Walter L. "Skip" Sharp, USA
Commander, United Nations Command
Commander, Republic of Korea-U.S. Combined Forces Command
Commander, U.S. Forces Korea

The full committee: Fiscal Year 2010 National Defense Authorization Budget Request from the Department of Defense, 13 May 2009.

Witnesses:
The Honorable Robert M. Gates
Secretary of Defense
Admiral Michael G. Mullen, USN
Chairman; Joint Chiefs of Staff

The full committee: Fiscal Year 2010 National Defense Authorization Budget Request from the Department of the Navy, 14 May 2009.

Witnesses:
The Honorable B.J. Penn
Acting Secretary of the Navy
Admiral Gary Roughead, USN
Chief of Naval Operations
General James T. Conway, USMC
Commandant of the Marine Corps

The full committee: Fiscal Year 2010 National Defense Authorization Budget Request from the Department of the Army, 14 May 2009.

Witnesses:
Honorable Pete Geren
Secretary of the Army
General George W. Casey, Jr., USA
Chief of Staff
U. S. Army

The Full Committee: The Fiscal Year 2010 National Defense Authorization Budget Request from the Department of the Air Force, 19 May 2009.
Witnesses:
The Honorable Michael B. Donley
Secretary of the Air Force
General Norman A. Schwartz, USAF
Chief of Staff
U.S. Air Force

The Military Personnel Subcommittee: The Fiscal Year 2010 National Defense Authorization Budget Request on Military Personnel Overview, 21 May 2009.

Witnesses:
Ms. Gail H. McGinn
Acting Under Secretary of Defense,
Personnel and Readiness
Department of Defense
Lieutenant General Michael D. Rochelle, USA
Deputy Chief of Staff, G-1
Headquarters, U.S. Army
Vice Admiral Mark E. Ferguson, III, USN
Chief of Naval Personnel
Deputy Chief of Naval Operations, Total Force
Lieutenant General Ronald S. Coleman, USMC
Deputy Commandant for Manpower and Reserve Affairs

Headquarters, U.S. Marine Corps
Lieutenant General Richard Y. Newton, III, USAF
Deputy Chief Staff, Manpower & Personnel
Headquarters, U.S. Air Force

The Terrorism, Unconventional Threats and Capabilities Subcommittee: Fiscal Year 2010 National Defense Authorization Budget Request for the U.S. Special Operations Command. 4 June 2009.

Witness:
Admiral Eric T. Olson, USN
Commander
U.S. Special Operations Command

The full committee met to receive testimony on the Fiscal Year 2011 National Defense Authorization Budget Request from the Department of Defense, 3 February 2010.

Witnesses:
The Honorable Robert M. Gates
Secretary of Defense
Admiral Michael G. Mullen, USN
Chairman
Joint Chiefs of Staff

The full committee met to receive testimony on the Fiscal Year 2011 National Defense Authorization Budget Request from the Department of the Air Force, 23 February 2010.

Witnesses:
The Honorable Michael B. Donley
Secretary of the Air Force
General Norton A. Schwartz, USAF
Chief of Staff
U.S. Air Force

The full committee met to receive testimony on the Fiscal Year 2011 National Defense Authorization Budget Request from the Department of the Navy, 24 February 2010.

Witnesses:
The Honorable Ray Mabus
Secretary of the Navy
Admiral Gary Roughead, USN
Chief of Naval Operations
U.S. Navy

General James T. Conway, USMC
Commandant
U.S. Marine Corps

The full committee met to receive testimony on the Fiscal Year 2011 National Defense Authorization Budget Request from the Department of the Army, 25 February 2010.

Witnesses:
The Honorable John McHugh
Secretary of the Army
General George W. Casey, Jr., USA
Chief of Staff
U.S. Army

The full committee met to receive testimony on the Fiscal Year 2011 National Defense Authorization Budget Request from the U.S. European Command, U.S. Africa Command, and U.S. Joint Forces Command, 10 March 2010.

Witnesses:
Admiral James G. Stavridis, USN
Commander
U.S. European Command
General William E. "Kip" Ward, USA
Commander
U.S. Africa Command
General James N. Mattis, USMC
Commander
U.S. Joint Forces Command

The Strategic Forces Subcommittee met to receive testimony on the status of United States Strategic Forces, 16 March 2010.

Witnesses:
General Kevin P. Chilton, USAF
Commander,
U.S. Strategic Command
Dr. James N. Miller
Principle Deputy Under Secretary of Defense for Policy
U.S. Department of Defense

The full committee met to receive testimony on the Fiscal Year 2011 National Defense Authorization Budget Request from the U.S. Central Command, U.S. Special Operations Command, and U.S. Transportation Command, 17 March 2010.

Witnesses:
General David H. Petraeus, USA
Commander
U.S. Central Command
Admiral Eric T. Olson, USN
Commander
U.S. Special Operations Command
General Duncan J. McNabb, USAF
Commander
U.S. Transportation Command

The full committee met to receive testimony on the Fiscal Year 2011 National Defense Authorization Budget Request from the U.S. Southern Command and U.S. Northern Command, 18 March 2010.

Witnesses:
General Douglas Fraser, USAF
Commander
U.S. Southern Command
General Victor E. Renuart, Jr., USAF
Commander
Northern American Aerospace Defense Command
U.S. Northern Command

The full committee met to receive testimony on the Fiscal Year 2011 National Defense Authorization Budget Request from the U.S. Pacific Command and U.S. Forces Korea, 25 March 2010.

Witnesses:
Admiral Robert F. Willard, USN
Commander
U.S. Pacific Command
General Walter L. "Skip" Sharp, USA
Commander
U.S. Forces Korea

The Strategic Forces Subcommittee met to receive testimony on the Fiscal Year 2011 National Defense Authorization Budget Request for Department of Energy atomic energy defense activities, 25 March 2010.

Witnesses:
The Honorable Thomas P. D'Agostino
Under Secretary for Nuclear Security and Administrator,
National Nuclear Security Administration
U.S. Department of Energy
The Honorable Inés R. Triay, Ph.D.

Assistant Secretary for Environmental Management
U.S. Department of Energy
The Honorable Peter S. Winokur, Ph.D.
Chairman,
Defense Nuclear Facilities Board
Defense Nuclear Safety Board

Member Meetings

HASC Chairman and Major General Robert M. Williams USAWC, 17 March 2009.
HASC Chairman and The Honorable William Navas, 18 March 2009.
"The Background and Basics of Professional Military Education,"
Dr. David Tretler, 24 March 2009.
Chairman, Ranking Member, and Lieutenant General Frances C. Wilson, 25 March 2009.
Chairman, Ranking Member, and Rear Admiral James P. Wisecup, 28 April 2009.
Chairman and Lieutenant General Allen G. Peck, 12 May 2009.
Chairman and ARNG USAWC Students, 13 May 2009.
Chairman, Ranking Member, and Admiral James Stavridis, 19 May 2009.
Chairman, Ranking Member, and Major General Robert P. Steel, 3 June 2009.
Chairman, Ranking Member, and Major General Robert M. Williams, 3 June 2009.
National War College Students, 4 June 2009.
Chairman, Ranking Member, and National War College Senior Leaders, 9 June 2009.
Chairman, Ranking Member, and General David Petraeus, 10 June 2009.
Chairman and Lieutenant General William B. Caldwell IV, 17 June 2009.
Chairman, Ranking Member, and General Kevin Chilton, 8 July 2009.
Chairman and Secretary of Defense Industry Fellows, 8 July 2009.
Chairman, Ranking Member, and Admiral Eric T. Olson, 14 July 2009.
Chairman, Ranking Member, and Vice Admiral Ann E. Rondeau, 23 July 2009.
Chairman, Ranking Member, and General James N. Mattis, 24 July 2009.
HASC Chairman and Advanced Operational Arts Studies Fellows, 28 July 2009.
Chairman, Ranking Member, and Lieutenant General Michael C. Gould,
15 September 2009.
Chairman, Ranking Member, and General Duncan McNabb, 15 September 2009.
HASC Chairman and General Martin Dempsey, 17 September 2009.
General James Conway, 24 September 2009.
Chairman with Congressional Research Service Staff, 25 September 2009.
Admiral Gary Roughead, 29 September 2009.
Chairman, Ranking Member, and General Victor Renuart, 30 September 2009.
General Norton Schwartz, 1 October 2009.
General George Casey, 21 October 2009.
Chairman, Ranking Member, and Admiral Michael Mullen, 27 October 2009.
Chairman, Ranking Member, and Major General Robert B. Neller, 5 November 2009.
Chairman, Ranking Member, and General William E. Ward, 5 November 2009.
Chairman and Vice Admiral Ann E. Rondeau, 16 March 2010.

Member Travel

Industrial College of the Armed Forces, 31 March 2009.
National Defense University and Naval War College, International Fellows Reception, 28 April 2009.
National War College, 29 April 2009.
Marine Corps University, 20 May 2009.
U.S. Naval Academy, 9 June 2009.
Marine Corps University, 4 March 2010.

Staff Meetings and Briefings

Rep. Solomon Ortiz (1989 Skelton Panel), 11 March 2009.
JCS/DJ-7, Briefing, 17 March 2009.
Office of the Secretary of Defense USD(P&R) and Defense Language Office Staffs, 20 March 2009.
U.S. JFCOM, *Joint Operating Environment (JOE)* Briefing to HASC Staff, 20 March 2009.
OSD, Briefing on PME, 20 March 2009.
Admiral Robert Natter (ret.) (1989 Skelton Panel Staff), 20 March 2009.
U.S. Marine Corps, PME Briefing, 23 March 2009.
U.S. Air Force, PME Briefing, 26 March 2009
U.S. Army, PME Briefing, 26 March 2009.
U.S. Navy, PME Briefing, 31 March 2009.
OSD and JCS, Quadrennial Defense Review Briefing to HASC Staff, 13 April 2009.
National Faculty Interviews, 14 April 2009.
JCS Staff, Joint Qualified Officer Management Briefing, 16 April 2009.
Dr. John Yaeger and Dr. Kjonnerod,
National Security Professional Development Program, 24 April 2009.
Rear Admiral James P. Wisecup, 28 April 2009.
National Faculty Interviews, 8 May 2009.
U.S. Navy (N15), PME Briefing, 11 May 2009.
Dr. Maren Leed, Center for Strategic and International Studies, 13 May 2009.
John Charlton, National Defense University International Fellows Briefing, 22 May 2009.
USAWC and NDU Faculty, On Teaching Strategy, 2 June 2009.
ICAF Faculty Interviews, 2 June 2009.
ICAF Faculty Interviews, 8 June 2009.
JCS/DJ-7, Joint Faculty Education Conference, 18-19 June 2009.
Former NDU Provost Interview, 22 June 2009.
NDU Staff, On Teaching Strategy, 22 June 2009.
Lieutenant General David Ohle (Ret.) and Lieutenant General Joseph E. DeFrancisco (Ret.), 1 July 2009.
Brigadier General Michael E. Rounds, JCS/DJ-7, 9 July 2009.
U.S. SOCOM, VTC Briefing on PME, 17 July 2009.

Brigadier General Edward Cardon, 17 July 2009.
USAWC Board of Visitors Members, 17 July 2009.
NDU Staff, 27 July 2009.
OSD and JCS, Quadrennial Defense Review Briefing to HASC Staff, 28 July 2009.
ICAF Senior Leaders, 29 July 2009.
Lieutenant General David Ohle (ret.) and Lieutenant General Joseph E. DeFrancisco (Ret.), 31 July 2009.
Vice Admiral Anne E. Rondeau, 7 August 2009.
Dr. Steven Randolph, Industrial College of the Armed Forces, 7 August 2009.
Colonel Mace Carpenter, Industrial College of the Armed Forces,
Draft Strategy Track and Ph.D. Program, 12 August 2009.
Brigadier General Edward Cardon, 17 August 2009.
Dr. Charles Stevenson, 17 August 2009.
Captain Frank Pagano (Ret.), NDU Capstone Briefing, 19 August 2009.
Services and Joint Staff meeting with HASC Staff on JPME I Faculty, 21 August 2009.
Dr. John Nagl, Center for a New American Security, 25 August 2009.
Captain Mark Hagerott, USN, 26 August 2009.
RAND, Congressional Roundtable on Military Personnel Issues, 27 August 2009.
Gail McGinn, USD(P&R), 27 August 2009.
USD (Policy) Briefing on the Guidance for the Development of the Force (Classified), 27 August 2009.
Lieutenant Colonel Paul Yingling, Telephone Interview, 27August 2009.
NDU Staff Interview, 28 August 2009.
Lieutenant General David W. Barno, USA (Ret.), 28 August 2009.
Tom Ricks, 8 September 2009.
U.S. JFCOM, Briefing, 9 September 2009.
NDU, Capstone Briefing, 9 September 2009.
NDU Provost, 14 September 2009.
Major General Robert Scales (Ret.), 14 September 2009.
Colonel John Collins (Ret.), 14 September 2009.
General Martin E. Dempsey, 17 September 2009.
ICAF Faculty Interview, 17 September 2009.
JCS Staff, Joint Officer Project (Fenty Study) Briefing, 17 September, 2009.
Brigadier General H.R. McMaster, 18 September 2009.
NWC Students, 23 September 2009.
Congressional Research Service Staff, 23 September 2009.
Dr. Albert Pierce, NDU, 28 September 2009.
USAFA Board of Visitors Meeting, 1 October 2009.
U.S. Army Staff FA 59 Briefing, 7 October 2009.
NDU President, 7 October 2009.
Rear Admiral Stephen R. Pietropaoli (Ret.), 9 October 2009.
Former Dean, National, 14 October 2009.
Former USAFA Faculty, 28 October 2009.
NDU President, 29 October 2009.
Representative Steve Israel, 4 November 2009.
JCS Staff, "JOM, JDAL, and JQS" Briefing, 6 November 2009.

ICAF Faculty, 24 November 2009.
NDU Ethics Center, 25 November 2009.
DUSD(P) Kathleen Hicks, 14 December 2009.
Dr. Andrew Krepinevich, 6 January 2010.
National Faculty, 22 January 2010.
Lieutenant General David Barno (Ret.), 22 January 2010.
General Richard B. Myers, 26 January 2010.
National Faculty, 29 January 2010.
Former Dean, National, 2 March 2010.
Lieutenant General David Barno (Ret.), 9 March 2010.
On Teaching Strategy Conference, USAWC and SSI, 9 April 2010.

Staff Travel

National Defense University: Lieutenant General Frances C. Wilson, Dr. John Deegan, Jr., and Dr. John W. Yaeger, 16 March 2009.
Marine Corps University, 27 March 2009.
National War College, 3 April 2009.
U.S. Army War College, Naval War College, and Air University, 5-9 April 2009.

U.S. Army Training and Doctrine Command, Joint Forces Staff College, and U.S. Joint Forces Command, 30 April-1 May 2009.
Joint Task Force-Bravo, U.S. Southern Command, and U.S. Central Command, 4-7 May.
U.S. Army Command and General Staff College, U.S. Transportation Command, and U.S. Strategic Command, 26-29 May 2009.
JCS Joint Faculty Education Conference, National Defense University, 18-19 June 2009.
U.S. Air Force Academy, U.S. Northern Command, and U.S. Military Academy, 29 June – 1 July 2009.
NDU, NDU President Interview, 10 July 2009.
Center for Strategic and International Studies, 23 September 2009.
NDU Leader Development Symposium, 8-9 October 2009.
ABA on PME, 27 January 2010.

APPENDIX E: BIBLIOGRAPHY

Government Documents

Academic Freedom Advisory Committee. *Academic Freedom Advisory Committee Report*, U.S. Military Academy, 22 May 2009.
Air Command and Staff College. "Self Study Resident Program." *Process for Accreditation of Joint Education*. Maxwell Air Force Base, September 2007.

________. "Self Study Non-Resident Program." *Process for Accreditation of Joint Education.* Maxwell Air Force Base, September 2007.

________. "Self Study Distance Education Program." *Process for Accreditation of Joint Education.* Maxwell Air Force Base, 22 August 2007.

Air University. "Professional Development (Advanced Academic Degrees and Professional Continuing Education)."

Air Force Instruction 36-2302, Air University Supplement. Maxwell Air Force Base, 4 June 2008.

________. "Academic Freedom." *Air University Instruction 36-2308.* Maxwell Air Force Base, 22 August 2008.

________. *Air University Faculty Handbook.* Maxwell Air Force Base, 2008.

Air University, Department of International Security Studies. *Regional and Cultural Studies Guide Book Academic Year 2009.* Maxwell Air Force Base, September 2008.

Air War College. "JPME Phase II Self-Study Report." Process for Accreditation of Joint Education, Maxwell Air Force Base, September 2008.

________. *Student and Faculty Handbook AY 2009.* Maxwell Air Force Base, 15 July 2008.

Booz Allen Hamilton. "Independent Study of Joint Officer Management and Joint Professional Military Education." McLean, 17 March 2003.

Catapult Consultants, LLC. *CAPSTONE Program Independent Review Final Report*, January 2009.

Congressional Research Service. "Department of Defense Reorganization Act of 1986: Proposals for Reforming the Joint Officer Personnel Management Program." *Congressional Research Service Report for Congress*, by Katherine Lemay Brown. Washington, D.C., 18 July 2000.

Chairman of the Joint Chiefs of Staff. "Defense Senior Leader Development Program." *Memorandum for Chiefs of Staff, the Commandant of the Marine Corps, Commanders of the Combatant Commands, and the President of National University*, by Gen. James E Cartwright, Acting Chairman of the Joint Chiefs of Staff. Washington, D.C., 27 June 2008.

________. "2009 Joint Professional Military Education (JPME) Special Areas of Emphasis (SAEs)." *Memorandum for Chief of Staff US Army, Chief of Naval Operations, Chief of Staff US Air Force, Commandant of the Marine Corps,* by Gen. James E Cartwright, Acting Chairman of the Joint Chiefs of Staff. Washington D.C., 23 April 2009.

________. "Officer Professional Military Education Policy (OPMEP)." *Instruction 1800.01C,* 22 December 2005, updated 7 August 2007.

________. "Officer Professional Military Education Policy (OPMEP)." *Instruction 1800.01D,* 15 July 2009.

________. "National Defense University Policy." *Instruction 1801.01A*, 1 January 2008.

________. "CJCS Vision for Joint Officer Development." *Memorandum for Chiefs of Staff,* November 2005.

________. "Defense Senior Leader Development Program (DSLDP)." *Memorandum for Chiefs of Staff,* by Gen. James E Cartwright, Acting Chairman of the Joint Chiefs of Staff. Washington, D.C., 27 June 2008.

________. "Capstone Course." *Memorandum for Service Chiefs of Staff and Commandants (CM 0796-09)* by Gen. James E Cartwright, Acting Chairman of the Joint Chiefs of Staff. Washington, D.C., 16 July 2009.

________. "Joint Professional Military Education Phase I (JPME I) Equivalent Credit." *Memorandum for Chiefs of Staff*, by ADM Michael G. Mullen, Chairman of the Joint Chiefs of Staff. Washington, D.C., 27 October 2008.

________. "Joint Education Courses Available through the Department of Defense in Calendar Year 2008." *Report to Congress*, 1 April 2009.

________. *Process for Accreditation of Joint Education (PAJE) US Air Force Command and Staff College*. Maxwell Air Force Base, 15-19 October 2007.

________. *Process for Accreditation of Joint Education (PAJE) USA Command and General Staff College*. Fort Leavenworth, 25-29 February 2008.

________. *Process for Accreditation of Joint Education (PAJE) College of Naval Command and Staff at the US Naval War College*. Newport, 5-10 October 2003.

________. *Process for Accreditation of Joint Education (PAJE) Joint and Combined Warfighting School*. Norfolk, 21-25 April 2008.

________. *Process for Accreditation of Joint Education (PAJE) Joint Advanced Warfighting School*. Norfolk, 21-25 April 2008.

________. *Process for Accreditation of Joint Education (PAJE) US Marine Corps Command and Staff College*. Quantico, 22-26 September 2008.

________. *Process for Accreditation of Joint Education (PAJE) US Air War College*. Maxwell Air Force Base, 20-23 October 2008.

________. *Process for Accreditation of Joint Education (PAJE) College of Naval Warfare*. Newport, 14-18 May 2007.

________. *Process for Accreditation of Joint Education (PAJE) College of Naval Warfare*. Newport, 4-7 May 2009

________. *Process for Accreditation of Joint Education (PAJE) Industrial College of the Armed Forces*. Washington, D.C., 6-11 February 2005.

________. *Process for Accreditation of Joint Education (PAJE) Marine Corps War College*. Quantico, 12-14 January 2009.

________. *Process for Accreditation of Joint Education (PAJE) National War College*. Washington, D.C., 14-19 November 2004.

________. *Process for Accreditation of Joint Education (PAJE) US Army War College*. Carlisle Barracks, 23-28 September 2007.

________. *Process for Accreditation of Joint Education (PAJE) US Army War College*. Carlisle Barracks, 21-25 September 2009.

________. *Capstone Concept for Joint Operations, Version 3.0*. 15 January 2009.

________. "Doctrine for the Armed Forces of the United States." *Joint Publication 1*, 2 May 2007, updated 20 March 2009.

________. "Department of Defense Dictionary of Military and Associated Terms." *Joint Publication 1-02*, 12 April 2001, updated 17 March 2009.

________. "Joint Intelligence." *Joint Publication 2-0*, 22 June 2007.

________. "Joint Operations." *Joint Publication 3-0*, 17 September 2006, updated 13 February 2008.

________. "Joint Logistics." *Joint Publication 4-0*, 18 July 2008.

________. "Joint Operation Planning." *Joint Publication 5-0*, 26 December 2006.

________. "Joint Communications System." *Joint Publication 6-0*, 20 March 2006.

________. "CAPSTONE Course Guidance." *Memorandum for the President, National Defense University*, 23 June 2009.

________. "Universal Joint Task List." *Chairman of the Joint Chiefs of Staff Manual 3500.04C*. Washington D.C., 1 July 2002.

________. "Language and Regional Expertise Planning." *Chairman of the Joint Chiefs of Staff Instruction 3126.01*. Washington, D.C., 23 January 2006, updated 11 February 2008.

________. "Joint Officer Management Program Procedures." *Chairman of the Joint Chiefs of Staff Instruction 1330.05*, 1 May 2008.

________. "Manpower and Personnel Actions Involving General and Flag Officers." *Chairman of the Joint Chiefs of Staff Instruction 1331.01C*, 22 July 2005, current as of 31 July 2007.

________. *The National Military Strategy of the United States of America*. Washington, D.C., 2004.

________. "2008 Joint Professional Military Education (JPME) Special Areas of Emphasis (SAEs)." *Memorandum for Chief of Staff US Army, Chief of Naval Operations, Chief of Staff, US Air Force, Commandant of the Marine Corps*, by ADM Michael G. Mullen, Chairman of the Joint Chiefs of Staff. Washington, D.C., 10 March 2008.

________. CM 344-90 *Military Education Policy Document*, by General Colin L. Powell, Chairman of the Joint Chiefs of Staff. Washington, D.C., 1 May 1990.

________. "Redelegation of Title 10 Hiring Authority." *Memorandum for the President, National Defense University*, by General Colin L. Powell, Chairman of the Joint Chiefs of Staff. Washington, D.C., 11 May 1990.

Congressional Research Service. "The DOD Service Academies: Issues for Congress." *Congressional Research Service Report for Congress*, by Robert L. Goldich. Washington, D.C., 6 February 1997.

________. "Legislative History of Joint Professional Military Education." *Memorandum to the House Armed Services Committee*, by Charles A. Henning. Washington D.C., 13 March 2009.

________. "Copyright Issues and Department of Defense Personnel." *Memorandum to House Armed Services Committee, Oversight and Investigations Subcommittee*, by Brian T. Yeh, 25 June 2009.

________. "Possible Approaches to Amending the Copyright Law to Allow Faculty of Military Service Academies and Department of Defense Professional Schools to Secure Copyrights for Their Scholarly Works." *Memorandum to House Armed Services Committee, Oversight and Investigations Subcommittee*, by Brian T. Yeh, 6 November 2009.

Defense Science Board. "Transition to and from Hostilities." *Defense Science Board 2004 Summer Study*. Washington, D.C., December 2004.

________. "Memorandum for Chairman, Defense Science Board Re: Terms of Reference, Defense Science Board Task Force on Joint Professional Military Education." Washington, D.C., 16 January 2009.

Deputy Secretary of Defense. "The Senior Service Colleges: Conclusions and Initiatives." *The Memorandum for Department of the Army, Department of the Navy, Department of the Air Force, the Joint Chiefs of Staff*. Washington, D.C., 5 June 1975. (Clements Report)

Education and Training Joint Cross-Service Group. *Base Closure and Realignment Report, Vol. VI*. Washington, D.C., May 2005.

Government Accountability Office. "Joint Officer Development has Improved, but a Strategic Approach is Needed." *Report to the Subcommittee on Military Personnel, Committee on Armed Services, House of Representatives (GAO-03-238).* Washington, D.C., December 2002.

Government Accountability Office (GAO), *Department of Defense: Professional Military Education at Three Senior Service Schools*, GAO/NSIAD-91-202, June 1991.

________. *Department of Defense: Professional Military Education at the Four Intermediate Service Schools*, GAO/NSIAD-91- 182, June 1991.

________. *Air Force: Status of Recommendations on Officers' Professional Military Education*, GAO/NSIAD-91-122BR, March 1991.

________. *Navy: Status of Recommendations on Officers' Professional Military Education*, GAO/NSIAD-91-124BR, March 1991.

________. *Army: Status of Recommendations on Officers' Professional Military Education*, GAO/NSIAD-91-121BR, March 1991.

________. *Marine Corps: Status of Recommendations on Officers' Professional Military Education*, GAO/NSIAD-91-88FS, February 1991.

________. *Military Education: Curriculum Changes at the Armed Forces Staff College*, GAO/NSIAD-91-288, September 1991.

________. *Military Education: Implementation of Recommendations at the Armed Forces Staff College*, GAO/NSIAD-92-30, October 1991.

________ . *Military Education: Implementation of Recommendations at the National War College*, GAO/NSIAD-92-202, June 1992.

________. *Military Education: Implementation of Recommendations at the Industrial College of the Armed Forces*, GAO/NSIAD- 92-221, June 1992.

________. *Military Education: Actions on Recommendations Involving Institute for Strategic Studies and Capstone*, GAO/NSIAD-92-280, September 1992.

________. *Military Education: DOD Needs to Enhance Performance Goals and Measures to Improve Oversight of Military Academies*, GAO-03-1000, September 1993.

________. "Military Education: DOD Needs to Develop Performance Goals and Metrics for Advanced Distributed Learning in Professional Military Education." *Report to the Ranking Minority Member, Committee on Armed Services, House of Representatives (GAO-04-873).* Washington, D.C., July 2004.

Hummel, Colonel Laurel J. "Memorandum for Chair, Assessment Steering Committee, United States Military Academy." 17 February 2010.

Industrial College of the Armed Forces. "Industrial College of the Armed Forces Organizational Chart." 16 March 2009.

________. "Student Evaluations/Grading Policy Academic Year 2008-2009." 2008-2009.

________. Syllabi (Compact Disk), 2008-2009.

Jackson, Tim. "PNWC/Provost Brief, Naval War College, College of Distance Education." November 2008.

The Joint Staff. "The Joint Staff Officer Project Final Report." *Prepared for Joint Staff J-7/Joint Training Division*, by Linda Fenty. April 2008.

________. "Assignment of Military Faculty at Service Senior Level Colleges." *(JPME II) Memorandum of Agreement*, 30 December 2005.

Larson, Admiral (Ret.) Charles R., *Study and Report Related to Permanent Professors at the U.S. Air Force Academy.* April 2004. Lynes, Colonel Jerome. "CJCS Vision for Joint

Officer Development: Roadmap." Military Education Coordination Council, November 2006.
Marine Corps University. *Title 10 Faculty Handbook*. Quantico, VA, June 2008.
Marine Corps University Institutional Research Assessment and Planning. *Marine Corps University Factbook 2008-2009*. Quantico, 2008.
Mattis, Gen. James. "Joint Forces Command Response to HASC O&I Report on Language Skills and Cultural Competencies in the Military." Norfolk, VA, 17 February 2009.
Mullen, ADM; Michael G. "National Defense University Commencement." Speech at the National Defense University, Washington, D.C., 11 June 2009.
________. "National Defense University: Change of Presidency." Speech at the National Defense University, Washington, D.C., 10 July 2009.
National Defense University. "2008 Annual Report." Washington, D.C., 1 January 2009.
________. "2009 Annual Report." Washington, D.C., revised March 2010.
________. Self-Study Design Plan Submitted to Middle States Commission on Higher Education (MSCHE). 15 March 2010.
________. "National Defense University President's Intent." *Memorandum for Senior Leaders*, by VADM Ann E. Rondeau. Washington, D.C., 25 August 2009.
________. "Report of the Board of Visitors Meeting." 2-3 December 2008.
________. "Vision, Mission Values, Strategic Goals." *Strategic Plan*, Washington, D.C., 15 May 2007.
________. "Personnel-Civilian." *National Defense University Regulation 690-4*. Washington, D.C., 4 August 2005.
________. "National Security Education Consortium, Pilot Program, AY 2008." Washington, D.C., 18 September 2008.
________. National War College Core Course Syllabi, 2008-2009.
________. Board of Visitors. "Minutes Of The National Defense University Board Of Visitors Meeting. 12-13 November 2009.
Naval War College. "Maritime Staff Operators Course, Syllabus Version 9.2." January 2009.
________. *Faculty Handbook*. October 2007.
________. "Faculty Diversity." *Statement of Academic Policy 09-01*, by Amb. Mary Ann Peters. Newport, 22 May 2009.
Office of the President of the United States. *The National Security Strategy of the United States of America*. Washington, D.C., 16 March 2006.
Office of the Secretary of Defense, Logistics and Materiel Readiness. "DOD Logistics Human Capital Strategy." 12 May 2008.
Office of the Under Secretary of Defense for Personnel and Readiness. "Report to Congress on Joint Officer Management." Washington, D.C., 30 April 2009.
RAND National Defense Research Institute. "A Framework for Joint Officer Management." *Research Brief*, 2005.
Spain, William. "Academics Brief Prepared for Rear Admiral James Wisecup, President of the Naval War College." 12 November 2008.
U.S. Air Force. "Professional Military Education." *Air Force Instruction 36-2301*, 27 June 2002.
________. "Commissioning Education Program." *Air Force Instruction 36-2014*, 22 April 2008.
________. "Total Force Development." *Air Force Policy Directive 36-26*, 27 August 2008.

________. "Executing Total Force Development." *Air Force Instruction 36-2640*, 16 December 2008.

________. "Foreign Language Requirements in the U.S. Air Force Academy Core Curriculum." Last revised 30 December 2004.

________. "Inspector General Activities." *Air Force Instruction 90-201,* 22 November 2004.

________. "Subject: Appointment to the Air Force Learning Committee (AFLC)." *Memorandum for Air Force Learning Committee Members*. Washington, D.C., 28 July 2008.

________. *Officer Force Development Panel Charter*, by Lt. Gen. Richard Y. Newton III and Duncan J. McNabb. 13 May 2008.

________. Responses to Officer Professional Military Education HASC Briefing Questions. 26 March 2009.

________. "Leadership and Force Development." *Air Force Doctrine Document 1-1*, 18 February 2006.

________. "Civilian Faculty Pay Plan for Air University and the USAF Academy." *Air Force Instruction 36-804*, 29 April 1994.

________. "Professional Military Education." *Air Force Instruction 36-2301*, 27 June 2002.

________. "Air Force Basic Doctrine." *Air Force Doctrine Document 1*, 17 November 2003.

________. "Air Force Glossary." *Air Force Doctrine Document 1-2*, 11 January 2007.

U.S. Air Force Academy Board of Visitors. "Semi-Annual Report, United States Air Force Academy Board of Visitors."

Memorandum for Secretary of Defense. Washington, D.C., 1 April 2009.

________. Minutes of The Regular Meeting Of The Board Of Visitors. 6-7 March 2009.

U.S. Air Force Academy. "Administering and Paying Civilian Faculty." *USAFA Instruction 36-802*, 26 April 2006, updated 21 May 2008.

________. "Appointment and Promotion to Academic Ranks." *USAFA Instruction 36-150*, 25 August 2009.

________. "Publication of Books and Other 'Works.'" *USAFA Instruction 51-303*, 22 November 2006.

________. "Intellectual Property--Patents, Patent Related Matters, Trademarks and Copyrights." *USAFA Instruction 51-303*, 1 September 1998.

________. "USAF Academy Speakers Program." *USAFA Instruction 36-3513*, 11 February 2009.

________. "U.S. Air Force Academy Diversity Plan." June 2009.

________. "50th Anniversary of USAFA." *Memorandum to For: Acting Dean and Faculty Council*, 26 August 2004.

________. "Clearance of Material for Public Release and Academic Freedom." *Faculty Operating Instruction 35-101*, 24 January 2008.

U.S. Army. "A Leader Development Strategy for an Expeditionary Army." 3 June 3009, updated 16 June 2009.

________. "Army Leadership." *Army Regulation 600-100*. Washington, D.C., 8 March 2007.

________. "Army Fellowships and Scholarships." *Army Regulation 621-7*. Washington, D.C., 8 August 1997.

________. "Army Training and Leader Development." *Army Regulation 350-1*. Washington, D.C., 3 August 2007.

________. "Commissioned Officer Professional Development and Career Management." *Pamphlet 600-3*. Washington, D.C., 11 December 2007.

________. "Leader Development for America's Army." *Pamphlet 350-58*. Washington, D.C., 13 October 1994.

________. "Army Leadership: Competent, Confident, and Agile." *FM 6-22 (FM 22-100)*. Washington, D.C., 12 October 2006.

________. "Organization and Functions, U.S. Army Training and Doctrine Command." *TRADOC Regulation 10-5*. Fort Monroe, 22 December 2005.

________. "The Army School System (TASS)." *TRADOC Regulation 350-18*. Fort Monroe, 3 April 2007.

________. "Institutional Leader Training and Education." *TRADOC Regulation 350-10*. Fort Monroe, 12 August 2002.

________. "Operations." *FM 3-0*. Washington, D.C., June 2001.

________. "Army Planning and Orders Production." *FM 5-0 (FM 101-5)*. Washington, D.C., January 2005.

________. "Mandatory Placements for Civilian Graduates of Senior Service Colleges (SSC)," by Gen. John M. Keane. Washington, D.C., 5 June 2003.

________. "The Army in Joint Operations: The Army's Future Force Capstone Concept 2015-2024 version 2.0."

TRADOC Pamphlet 525-3-0. Fort Monroe, 7 April 2005.

________. "Commissioned Officer Professional Development and Career Management." *Pamphlet 600-3*. Washington, D.C., 11 December 2007.

________. "Morale, Welfare, and Recreation: Leadership Development—Mentoring, Coaching and Counseling" U.S. Army Forces Command R*egulation 215-8*. Fort McPherson, 15 August 1999.

U.S. Army Command and General Staff College (CGSC). "Intermediate Level Education (ILE) Self Study." Fort Leavenworth, 8 January 2008.

U.S. Army War College. "Electives: Academic Year 2009." Carlisle Barracks, PA, 2009.

________. *Title 10 Civilian Faculty Manual CGSC Pamphlet 690-1*, Carlisle Barracks, PA, August 2008.

U.S. Central Command. "COS Indoctrination Training Course -- (Overview and Objectives)." *Syllabus*, 7 May 2009.

________. "Command Policy Letter Number 40." *Memorandum to US Central Command Headquarters Personnel*, by Maj. Gen. Jay W. Hood. MacDill Air Force Base, 9 April 2009.

U.S. Congress. *Title 10 U.S.C. Section 1595*, "Civilian faculty members at certain Department of Defense schools: employment and compensation."

________. *Title 10 U.S.C. Section 4021*, "Army War College and United States Army Command and General Staff College: civilian faculty members."

________. *Title 10 U.S.C. Section 7478*, "Naval War College and Marine Corps University: civilian faculty members."

________. *Title 10 U.S.C. Section 9021*, "Air University: civilian faculty members."

U.S. Congress. House of Representatives. Committee on Armed Services. *Report of the Panel on Military Education of the 100th Congress*. 100th Cong, 1st sess., 21 April 1989. (Skelton Report)

________. House of Representatives. Committee on Armed Services Military Education Panel, Testimony by John M. Collins. 100th Cong., 2nd sess., 21 April 1988.

________. House of Representatives. Committee on Armed Services, Subcommittee on Total Force. "A Strategic Approach is Needed to Improve Joint Officer Development," Testimony by Derek Stewart. 108th Cong., 1st sess. 19 March 2003.

U.S. Department of Defense. "Assignment of Military Faculty at Service Senior Level Colleges." *Memorandum of Agreement*, 30 December 2005.

________. "Civilian Faculty Positions in Department of Defense (DoD) Post-Secondary Educational Institutions." *Instruction Number 1402.06*, 6 November 2007.

________. "Joint Officer Management: Joint Qualification System Implementation Plan." 30 March 2007.

________. "Strategic Plan for Joint Officer Management and Joint Professional Military Education." 3 April 2006.

________. "DoD Joint Officer Management Program." *Instruction 1300.19*, 31 October 2007, updated 21 August 2008.

________. "Fellowships, Scholarships, Training With Industry (TWI), and Grants for DoD Personnel." *Instruction 1322.06*, 15 November 2007.

________. "Policy on Graduate Education for Military Officers." *Instruction 1322.10*, 19 April 2008.

________. *National Defense Strategy*. June 2008.

________. *Quadrennial Defense Review Report*. Washington, D.C., 6 February 2006.

________. *Quadrennial Defense Review Report*. Washington, D.C., February 2010.

________. "Service Academies." *Department of Defense Directive 1322.22*. 24 August 1994.

________. "Management of DoD Language and Regional Proficiency Capabilities." *Instruction 5160.70*, 12 June 2007.

________. "Clearance of DoD Information for Public Release." *Department of Defense Directive 5230.09*, 22 August 2008.

________. "Irregular Warfare (IW)." *Department of Defense Directive 3000.07*, 1 December 2008.

________. "Office of the Chancellor for Education and Professional Development." *Department of Defense Directive 5124.7*, 27 September 1999.

________. "DoD Joint Officer Management Program Procedures." *Department of Defense Instruction 1300.20*, 20 December 1996.

________. "Terms of Reference Fact Sheet," *2010 Quadrennial Defense Review*, 27 April 2009.

________. "Functions of the Department of Defense and Its Major Components." *Department of Defense Directive 5100.01*, 21 November 2003.

________. "Under Secretary of Defense for Personnel and Readiness." *Department of Defense Directive 5124.02*, 23 June 2008.

________. Legislative Proposal, "Sec. . Expansion Of Authority Relating To Phase II Of Three–Phase Approach To Joint Professional Military Education, 6 March 2010.

________. "The Armed Forces Officer." January 2006.

U.S. Government. *Counterinsurgency Guide*. January 2009.

U.S. Joint Forces Command. "Chairman's Statutory Responsibilities and Officer Joint Professional Military Education (JPME)." February 2009.

________. *The Joint Operating Environment 2008. Challenges and Implications for the Future Joint Force.* 25 November 2008.

U.S. Marine Corps. "Subject: Professional Military Education." *Marine Corps Order 1553.4B*. Washington, D.C., 25 January 2008.

________. "Vision and Strategy 2025 Implementation Planning Guidance."

________. "Marine Corps Operations." *Marine Corps Doctrinal Publication 1-0.* Washington, D.C., 27 September 2001.

________. "Warfighting." *Marine Corps Doctrinal Publication 1.* Washington, D.C., 20 June 1997.

U.S. Merchant Marine Academy. "Conference on Professional Military Education Rebuilding America's Intellectual Arsenal," *Report on the Proceedings of Conference convened by Congressman Steve Israel*, New York, 25 March 2006.

U.S. Military Academy. *Faculty Manual.* West Point, December 2005.

U.S. Military Academy. "Institutional Self Study Report." *Report of the United States Military Academy*, June 2009.

________. *Faculty Manual.* West Point, December 2005.

U.S. Naval Academy. "Life and Work at the United States Naval Academy." *Information for Faculty*, March 2008.

U.S. Navy. "Naval Warfare." *Naval Doctrine Publication 1.* Washington, D.C., 28 March 1994.

Wilhelm, General Charles E. (Ret.), Lieutenant General Wallas C. Gregson, Jr. (Ret.), Lieutenant General Bruce B. Knutson, Jr. (Ret), Lieutenant General Paul K. Van Riper (Ret.), Dr. Andrew F. Krepinevich, Jr., and Dr. Williamson Murray. *U.S. Marine Corps Officer Professional Military Education 2006 Study and Findings,* (Wilhelm Report). Marine Corps University, 29 September 2006.

Legislation

National Defense Authorization Act for Fiscal Years 1988 and 1989, P.L. 100-180

National Defense Authorization Act for Fiscal Year 1994, House Conference Report 103-357

National Defense Authorization Act for Fiscal Year 1996, P.L. 104-106

National Defense Authorization Act for Fiscal Year 2005, House Conference Report 108-767, 8 October 2004.

National Defense Authorization Act for Fiscal Year 2005, House Report 108-491, 14 May 2004.

National Defense Authorization Act for Fiscal Year 2007, Conference Report 109-702 National Defense Authorization Act for Fiscal Year 2007, Senate Report 109-254

National Defense Authorization Act for Fiscal Year 2010, House Conference Report 111-288, 7 October 2009.

Non-Government Documents

Army Times. "Perot pledges $6.1m to Leavenworth institution." 6 November 2009.

Associated Press. "U.S. Adm. James Stavridis Named Military Commander Of NATO." 13 May 2009.

Baker III, Fred W. "Mullen Offers Leadership Lesson to National Defense University Grads." *American Forces Press Service*, 11 June 2009.

Barno, David W. (LTG) (Ret.). "Senior Leader Development – Time for a 'College for Generals'" (Draft).

Bensahel, Nora; Olga, Oliker; Heather, Peterson. "Improving Capacity for Stabilization and Reconstruction Operations." *RAND National Defense Research Institute*, 2009.

Biddle, Stephen; Jeffrey, A. Friedman. "The 2006 Lebanon Campaign and the Future of Warfare: Implications for Army and Defense Policy." *Strategic Studies Institute*, September 2008.

Breslin-Smith, Janet. "A School for Strategy: The Early Years." *Joint Forces Quarterly,* Issue 41, 2nd Quarter, 2006.

Breslin-Smith, Janet; Cliff, Krieger. "Strategic Drift? The Future of the National War College." *Joint Forces Quarterly*, Issue 55, 4th Quarter, 2009.

Breslin-Smith, Janet. "Notes on the Future of National War College." Undated.

Brewster, Major John A. "Time to Overhaul the United States Air Force's Air Command and Staff College." *Air Command and Staff College, Air University*, Maxwell Air Force Base, Alabama, April 2006.

Brodie, Bernard. *War and Politics*. New York: Macmillan, 1973.

Carafano, James Jay; Alane, Kochems. "Rethinking Professional Military Education." *Executive Memorandum*, No. 976. The Heritage Foundation, 28 July 2005.

Chandrasekaran, Rajiv. "Pentagon Worries Led to Command Change." *Washington Post*, 17 August 2009.

Cheney, Richard, Bill Taylor. "Professional Military Education: An Asset for Peace and Progress." *A Report of the CSIS Study Group on Professional Military Education*, The Center for Strategic & International Studies, March 1997.

Clausewitz, Carl von. *On War*, trans. and ed. by Michael Howard and Peter Peret. Princeton: Princeton University Press, 1976.

Cochran, Alexander S. "American's War Colleges, Teaching the Right Stuff for the Decade Ahead: A Preliminary Report." Briefing presented to Woodrow Wilson Center for International Scholars, May 2005.

________. "War Colleges for the Next Decade: The Way Ahead Observations." 18 May 2009. Cole, Reyes. "The Military Diplomat." *Armed Forces Journal,* February 2009.

Coll, Steve. "The Future of Soldering." The New Yorker Online. (http://www.newyorker.com/online/blogs/stevecoll/2009/06/the-future-of-soldiering.html), 1 June 2009. Corbett, Julian S. *Some Principles of Maritime Strategy*. New York: Longmans, Green & Co., 1911.

da Costa, Thomaz Guedes. "The Teaching of Strategy and Exploring the Nexus Among Definition, Competences, and Curricula." Paper presented at the XX Annual Teaching Strategy Conference, U.S. Army War College, Carlisle, 17 April 2009.

De Vise, Daniel. "Naval Academy defends move to adjust image on national TV Membership of color guard at World Series was revised to reflect school's diversity." *The Washington Post* (on line), 10 November 2009.

Deegan, John, Jr. "First Step to Academic Excellence: The Faculty." 14 April 2009.

________. "Building a High Quality University." Remarks before the National Defense University Board of Visitors, 2 December 2008.

DePalma, Jr.; Thomas, J; William, D. Brobst. *Does a Technical Education Improve Tactical Performance?* Alexandria, Center for Naval Analysis, 2008.

Ewing, Philip. "Naval Academy has lower standards for racial minorities: professor." *Chicago Sun-Times*, 22 June 2009. Evaluation Team Representing the Middle States Commission on Higher Education. "Final Report to the Faculty, Administration, Trustees, Cadets of the United States Military Academy, 13-16 September 2009.

Fagin, Barry S; James, E. Parco. "Higher Power: Is Evangelism Still in Command at the U.S. Air Force Academy?" *The Humanist*, September/October 2007

Franke, Volker. "Making Sense of Chaos: Teaching Strategy Using Case Studies." Paper presented at the XX Annual Teaching Strategy Conference, U.S. Army War College, Carlisle, 17 April 2009.

Fuller, JFC. *Generalship: Its Diseases and Their Cure.* Harrisburg: Military Service Publishing Company, 1936. Gardner, Howard. "How Education Changes: Considerations of History, Science, and Values."

Galvin, John R. "What's the Matter with Being a Strategist?" *Parameters: U.S. Army War College Quarterly,* 19 March 1989.

Garvin, David A. "Making the Case: Professional education for the world of practice." *Harvard Magazine*, Vol. 6 No. 1, September-October 2003.

Gates, Robert M. "The National Defense Strategy, Striking the Right Balance." *Joint Forces Quarterly,* Issue 52, 1st Quarter, 2009.

Gray, Colin S. "Schools for Strategy: Teaching Strategy for 21st Century Conflict." *Strategic Studies Institute*, 28 October 2009.

Greer, Steven J. "Irregular Warfare Education: Winning the Long War." *American Chronicle*, 23 July 2009. (available online at http://www.americanchronicle.com/articles/view/ 111387).

Grice, LtCol Michael D. "Where Is Our Kilcullen: Professional relevance as a result of education." *Marine Corps Gazette*, December 2009.

Halsey, Ashley III. "Charting morally murky waters: midshipmen study ethical dilemmas in interactive DVDs." *The Washington Post*, 4 December 2009.

Hanser, Lawrence M; Louis W. Miller; Herbert J. Shukiar; Bruce Newsome. "Developing Senior Navy Leaders, Requirements for Flag Officer Expertise Today and in the Future." RAND National Defense Research Institute, 2008.

Harrell, Margaret C; Harry, J; Thie, Sheila N; Kirby, Al Crego; Danielle M. Varda; Thomas Sullivan. *A Strategic Approach to Joint Officer Management: Analysis and Modeling Results.* RAND National Defense Research Institute, 2009.

Harrell, Margaret C; John, F. Schank; Harry, J. Thie; Clifford, M; Graff, II; Paul, Steinberg. *How Many Can Be Joint? Supporting Joint Duty Assignments*. RAND National Defense Research Institute, 1996.

Harrison, Ross. "Teaching Strategy in 3D." (Paper presented at the XX Annual Teaching Strategy Conference, U.S. Army War College, Carlisle, 17 April 2009).

Hart, BH. Liddell. *Strategy: The Indirect Approach*. London: Faber and Faber, 1954. Hart, Kim. "Cyber Threats 101." *Washington Post*, 16 February 2009.

The Higher Learning Commission. "Assurance Section." *Report of Comprehensive Evaluation Visit*. 27-29 April 2009.

________. "Advancement Section." *Report of Comprehensive Evaluation Visit*. 27-29 April 2009.

Hill, Michael. "Cadet Training Has New Dimension: Army Tackles Ethics, Politics in Role-Playing." *The Washington Times*, 3 August 2009. (available online at http://www.washingtontimes.com/news/2009/aug/03/cadet-training-has- new-dimension/)

Hoffman, Frank G. "Hybrid Warfare and Challenges," *Joint Forces Quarterly*, Issue 52, 1st Quarter, 2009.

Holder, Leonard D. Jr; Williamson, Murray. "Prospects for Military Education." *Joint Forces Quarterly*, Spring 1998. Iklé, Fred Charles. "The Role of Character and Intellect in Strategy." In *On Not Confusing Ourselves: Essays on National Security Strategy in Honor of Albert and Roberta Wohlstetter,* eds. Andrew W. Marshall, J.J. Martin, and Henry S. Rowen. Westview Press: Boulder, CO, 1991.

Israel, Steve. "Commencement Address." Speech given for the Intermediate Level Education Graduation Class 08-01, location unknown, 13 June 2008.

Joint Forces Quarterly. Washington, DC: National Defense University Press. Issue 54, 3d Quarter 2009.

Jones, Matthew. "Experts' Vision Of Future Warfare Is Not Black And White." *The Virginian-Pilot*, 13 May 2009. Kabel, Marcus. "Civil War Guerrillas Officer Lessons for Army in Iraq." *The Associated Press*. 17 July 2006. Kennedy, Gregory C. and Keith Neilson. *Military Education: Past, Present, and Future*. Praeger: Westport, CT, 2002.

Kenney, Steven H. "Professional Military Education and the Emerging Revolution in Military Affairs." *Airpower Journal*, Fall 1996.

Kennedy, Robert. "The Elements of Strategic Thinking: A Practical Guide." Paper presented at the XX Annual Teaching Strategy Conference, US Army War College, Carlisle, 17 April 2009.

Kipp, Jacob W. "Moscow Examines an Alternative Vision of Military Professionalism." *Eurasia Daily Monitor*, Vol. 6, Issue 184, 7 October 2009.

Kirby, Sheila Nataraj; Al, Crego; Harry, Jl, Thie; Margaret, C; Harrell, Kimberly Curry; Michael, S Tseng. "Who Is 'Joint'? New Evidence from the 2005 Joint Officer Management Census Survey." RAND Corporation, National Defense Research Institute, 2006.

Kohn, Richard H. "Tarnished Brass: Is the U.S. Military Profession in Decline?" *World Affairs Journal*, Spring Edition, 2009.

Krepinevich, Andrew. *Memorandum to the Honorable Vic Snyder Re: Strategy and Professional Military Education*. 5 May 2009. Krepinevich, Andrew. *Rebuilding America's Intellectual Arsenal*. 18 December 2005.

Leed, Maren; David, Sokolow. *The Ingenuity Gap: Officer Management for the 21st Century*. Center for Strategic and Interdisciplinary Studies, January 2010.

Lorenz, LtGen Stephen R. "Transforming Air Force Education for the Long War and Beyond." *Air & Space Power Journal*, Summer 2007.

Mahoney-Norris, Kathleen. "*The Role Of Professional Military Ecication In Preparing Officers To Wage War Justly And Ethically*." Paper Prepared for Presentation at the 50th

Annual International Studies Association Convention, New York City, NY, 15-18 February 2009.

Marcella, Gabriel, Stephen O. Fought. "*Teaching Strategy in the Twenty First Century.*" Paper presented at the XX Annual Teaching Strategy Conference, US Army War College, Carlisle, 17 April 2009.

Marshall, Andrew W. "Commentary: Strategy as a Profession in the Future Security Environment." Revised and updated version of Marshall's essay, "Strategy as a Profession for Future Generations." In Marshall, J. J. Martin and Elenry S. Rowen eds., *On Not Confusing Ourselves: Essays on National Security Strategy in Honor of Albert and Roberta Wohstetter,*. Westview Press: Boulder, CO, 1991.

Mattis, General James. "Joint Warfare in the 21st Century," March 2009.

________. "Developing & Retaining The Officers We Need For The 21st Century." 18 February 2010.

Maxwell, David S. "A National Security Act of 2009? A Short Recommendation for a Possible Revision of the National Security Act of 1947." *Small Wars Journal*, 2009.

McCausland, Jeffrey D. "Developing Strategic Leaders for the 21st Century." *Strategic Studies Institute*, February 2008. McCausland, Jeffrey D. and Neil Weissman, *Educating Leaders in an Age of Uncertainty – The Future of Military War Colleges: A Research Study for the Smith Richardson Foundation*, Dickinson College Leadership in Conflict Initiative, Carlisle, PA, 15 December 2005.

McGuire, Mark A. "Senior Officers and Strategic Leader Development." *Joint Forces Quarterly,* Autumn/Winter, 2001. McMichael, William H. "'We need to build strategic leaders:' Report calls for new system for officer development." *Navy Times*, 1 March 2010.

McMichael, William H. "Gates Kicks Off Tour Of Military War Colleges." *Army Times*, 13 April 2009.

Micewski, Edwin R; Hubert, Annen (eds.). "Studies for Military Pedagogy, Military Science & Security Policy: Military Ethics in Professional Military Education – Revisited." New York: Peter Lang, 2005.

Moore, Craig S; Marygail, K. Brauner. "Advancing the US Air Force's Force-Development Initiative." RAND Project Air Force, 2007.

Moskowitz, Michael J; David, M. Rodney; Kletus, S. Lawler. *Data Analysis for a Navy Education Strategy*. Alexandria, Center for Naval Analysis, 2008.

Nagl, John A; Brian, M. Burton. "Keeping The Edge: Revitalizing America's Military Officer Corps." *Center for a New American Security*, February 2010.

________. "Keeping The Edge: Officership for a New Era." Center for a New American Security Officership Capstone 1st Draft, 28 August 2009.

Noonan, John. "The Thayer System, After Two Centuries of Success, It might be Time to Make Some Changes." *The Weekly Standard*, 1 August 2007.

Nye, Joseph S. Jr. "Scholars on the Sidelines." *The Washington Post*, 13 April 2009.

U.S. Army War College. *Parameters*. Carlisle, PA: U.S. Army War College. Vol. XXXVII, No. 1, Spring 2007.

Park, Francis J.H. "The Strategic Plans and Policy Officer in the Modular Division." *Military Review,* November- December 2007.

Paschal, David G. "Irregular Warfare: Impact on Future Professional Military Education." U.S. Army War College, Carlisle, 15 March 2006.

Perry, David L. "How Ethics is Taught at the U.S. Army War College," Edwin R. Micewski and Hubert Annen, eds., *Studies for Military Pedagogy, Military Science and Security Policy: Military Ethics in Professional Military Education – Revisited*, Peter Lang GmbH, Europaeischer Verlag der Wissenschaften, Frankfurt am Main, 2005.

Peterson, Laura. "Don't Know Much about Strategy. Is the Military shortchanging its war colleges – and its future?" *Boston Globe*, 9 July 2006.

Peterson, Zachary M. "283 Officers Seek New 'Joint' Status." *Navy Times*, 11 February 2008. Petraeus, General David H. "Beyond the Cloister." *The American Interest*, July/August 2007. Pierce, Albert C. "Academic Community." (Draft) 27 May 2009.

________. "Ethics Education in U.S. Professional Military Education." Remarks prepared for delivery at MEEN Conference, 10 December 2008.

Quenqua, Douglas. "With Enough Good Soldiers, the Army is Looking for a Few Good Officers." *New York Times*, 2 August 2009. (available online at http://www.nytimes.com/2009/08/03/business/media/03adco.html).

Record, Jeffrey. "The Creeping Irrelevance of U.S. Force Planning," Strategic Studies Institute, U.S. Army War College, Carlisle, 19 May 1998.

________. "Bounding the Global War on Terrorism," Strategic Studies Institute, U.S. Army War College, Carlisle, December 2003.

Reveron, Derek S; James, L. Cook. "Developing Strategists: Translating National Strategy into Theater Strategy." *Joint Forces Quarterly,* Issue 55, 4th Quarter, 2009.

Richardson, David K. "The Case for JPME Phase Zero: Building a Joint Culture in the U.S. Navy." *Joint Forces Quarterly*, Issue 50, 3rd Quarter, 1 July 2008.

Ricks, Thomas E. "Why We Should Get Rid of West Point." *Washington Post*, 19 April 2009.

________. "Outlook: Shut Down West Point, Annapolis, Air Force Academy: How to Trim the Federal Budget and Improve Leadership." *Washington Post on-line discussion*, 20 April 2009.

Rodney, David M; Christine H. Fox; Samuel D. Klienman; Michael J. Moskowitz; Mary E. Lauer. *Developing an Education Strategy for URL Officers.* Center for Naval Analysis, Alexandria, 2008.

Ruby, Tom. "Flying High, Thinking Big: Anti-intellectualism in the U.S. Air Force threatens its future, not ours." *The American Interest,* Summer Vol. IV, No. 5, May/June 2009.

Thornhill, Paula G. "Improving Capstone: Change course focus to challenge participants." *Armed Forces Journal (online edition)*, April 2010.

Scales, Major General Robert H. "Too Busy To Learn." *Proceedings*, February 2010.

________. *The Past and Present as Prologue: A View of Future Warfare Through the Lens of Contemporary Conflicts*. Center for a New American Security, 9 April 2009.

________. "Return of the Jedi." *Armed Forces Journal*, October 2009.

Schank, John F; Harry, J. Thie; Jennifer, Kawata; Margaret, C. Harrell; Clifford, M. Graf II; Paul, Steinberg. *Who is Joint? Reevaluatingthe Joint Duty Assignment List*. RAND National Defense Research Institute, 1996.

Schirmer, Peter; Harry, J. Thie; Margaret, C. Harrell; Michael, S. Tseng. *Challenging Time in DOPMA: Flexible and Contemporary Military Officer Management*. RAND National Defense Research Institute, 2006.

Schirmer, Peter; Dina, G. Levy; Harry, J. Thie; Joy, S. Moini; Margaret, C. Harrell; Kimberly, Curry; Kevin, Brancato; Megan, Abbott. *New Path to Success: Determining*

Career Alternatives for Field-Grade Officers. RAND National Defense Research Institute, 2004.

Schirmer, Peter. *Computer Simulation of General and Flag Officer Management: Model Description and Results*. RAND National Defense Research Institute, 2009.

Schlesinger, et al. "Final Report of the Independent Panel To Review DoD Detention Operations." August 2004. (Schlesinger Report).

Skelton, Ike. "Beyond Iraq." Speech given at the Dwight D. Eisenhower National Security Conference, Washington D.C., 28 September 2005.

________. "Remarks." *Speech at the Graduation Exercises of the Naval War College*. Newport, 19 June 2009.

________. *Whispers of Warriors: Essays on the New Joint Era*. Washington D.C.: National Defense University Press, 2004.

Stavridis, Admiral James; Captain, Mark Hagerott. "The Heart of an Officer: Joint, Interagency, and International Operations and Navy Career Development." *Naval War College Review,* Spring 2009, Vol. 62, No. 2, 2009.

Stiehm, Judith Hicks. *The U.S. Army War College: Military Education in a Democracy*. Temple University Press: Philadelphia, PA, 2002.

Stefanick, Michelle L. "Powell's Leadership Principles – Time for State Department to Revisit." U.S. Army War College, Carlisle, PA, 20 January 2009.

Stevenson, Charles A. *Congress at War. The Politics of Conflict since 1789*. National Defense University Press, Dulles, VA, 2007.

Thie, Harry J; Margarett, C; Harrell, Roland J; Yardley, Marian Oshiro; Holly, Ann Potter. *Framing a Strategic Approach for Joint Officer Management*. RAND National Defense Research Institute, 2005.

Thornhill, Paula G. "Flag Officer Education: It's Time to Improve the Capstone Course," *Armed Forces Journal* (online edition), April 2010.

Vego, Milan. "There's No Place Like Newport." *Proceedings*, February 2010.

Vlahos, Michael. *The Blue Sword: The Naval War College and the American Mission*. Newport: Naval War College Press, 1980. Wardynski, Casey, David S. Lyle, and Michael J. Colarusso. "Towards A U.S. Army Officer Corps Strategy for Success: A Proposed Human Capital Model Focused Upon Talent." Strategic Studies Institute, April 2009.

Wass, de Czege; Brigadier, General Huba (Ret.). "The School of Advanced Military Studies: An Accident of History." *Military Review*, July/August 2009.

Watson, Cynthia. *Military Education: A Reference Handbook*. Westport, CT: Praeger Security International, 2007.

Watts, Barry D. "U.S. Combat Training, Operational Art, and Strategic Competence." Center for Strategic and Budgetary Assessments, 2008.

Wehner, Peter. "How the U.S. Army Evolved." *Commentary Magazine Contentions Blog*, posted 18 August 2009. (available online at http://www.commentarymagazine.com/blogs/index.php/wehner/76372).

Wertheim, Mitzi. "Thinking About…Learning in DOD." Presented at Center for Naval Analysis Corporation, 21 June 2005.

Wiarda, Howard J. "*Comparative Politics, International Relations, and Foreign Policy*."

Wiite, Brian. "Military Academies see applications jump as tough economy pairs with stepped up recruiting." *Baltimore Sun*, 16 June 2009. *Issues in Foreign Policy, Comparative Politics and International Affairs,* Volume II, Number 7, (March 2009).

Yaeger, John W. "Congressional Influence on National Defense University." *Dissertation for Graduate School of Education and Human Development of The George Washington University*, 21 May 2005.

________."Congressional Influence on National Defense University." Volume II Appendixes. *Dissertation for Graduate School of Education and Human Development of The George Washington University*, 21 May 2005

________. "Developing National Security Professionals." *Joint Force Quarterly,* Issue 49, 2nd Quarter, 2008, 115-120.

________. "The Origins of Joint Professional Military Education." *Joint Forces Quarterly,* Issue 37, 2nd Quarter, 2005.

Yarger, Harry R. "How Do Students Learn Strategy? Thoughts of the USAWC Pedagogy of Strategy." Paper presented at the XX Annual Teaching Strategy Conference, U.S. Army War College, Carlisle, PA, 17 April 2009.

Yingling, Paul. "A Failure of Generalship." *Armed Forces Journal*, May 2007.

Briefings

Belcher, Col. Michael. "Marine Corps War College Curriculum Review Board (CRB)." Presented to House Armed Services Subcommittee on Oversight and Investigations Staff, 4 February 2009.

Bonner, Matthew. "Military Education Coordination Council 2009 Special Areas of Emphasis Areas Assessment." Presented to House Armed Services Subcommittee on Oversight and Investigations Staff, 12 February 2009.

Broadwell, Paula (MAJ). "Junior Officer Leadership A "Major Crisis?" Presented at the Marine Corps University symposium titled, "Counterinsurgency Leadership in Afghanistan, Iraq, and Beyond." National Press Club, Washington, D.C., 23 September 2009.

Pagano, Frank. "Capstone Introduction." Presented to House Armed Services Subcommittee on Oversight and Investigations, 19 August 2009.

Carpenter, Mace. "School of Advanced Strategic Planning and Strategy (ASPS)." Presented to House Armed Services Subcommittee on Oversight and Investigations, 9 June 2009.

Chairman of the Joint Chiefs of Staff. "CJCS Responsibilities by Law." Presented by JCS/DJ-7 to House Armed Services Subcommittee on Oversight and Investigations, 19 February 2009.

"Cognitive Transformation Roundtable." United States Capitol, Washington, DC. 16 February 2005.

Defense Language Office. "Regional Area Content in Officer Professional Military Education." Presented to House Armed Services Subcommittee on Oversight and Investigations Staff, 20 March 2009.

Garza, Brigadier General David. "Strategic Planning." *U.S. Southern Command*, Marine Corps War College, 8 December 2008.

Industrial College of the Armed Forces. "Resourcing the National Strategy." Presented to House Armed Services Subcommittee on Oversight and Investigations, 31 March 2009.

Joint Chiefs of Staff. (2009). "Active (AC) and Reserve (RC) Component By Comparison," undated. Kjonnerod, Erik L. "'We Live in Exponential Times' Interagency to Whole-of-Government."

Knarr, William. "Applying Current Wars' Lessons to Training and Education." Presented to House Armed Services Subcommittee on Oversight and Investigations, 12 February 2009.

________. "International Officers at JPME Schools, Maximizing Engagement Opportunities." Presented to House Armed Services Subcommittee on Oversight and Investigations, 12 February 2009.

________. "Process for Accreditation of Joint Education (PAJE) Analysis of PAJE Trends." Presented to House Armed Services Subcommittee on Oversight and Investigations, 12 February 2009.

Lynes, Jerome M. "Professional Military Education." Presented to House Armed Services Oversight and Investigations Subcommittee, 17 March 2009.

________. "Welcome Military Education Coordination Council." Presented to House Armed Services Subcommittee on Oversight and Investigations, 12 February 2009.

Mahar, Brian. "Update Briefing to the Military Education Coordination Council." Presented by Joint Special Operations University to House Armed Services Subcommittee on Oversight and Investigations, 12 February 2009.

Marine Corps University. "Congressional Delegation Discussion." Presented to House Armed Services Subcommittee on Oversight and Investigations, 20 May 2009.

________. "Congressional Delegation Visit." Presented to House Armed Services Subcommittee on Oversight and Investigations, 23 March 2009.

Marine Corps War College. (2008) "Leadership Learning Outcomes," Slide,

Military Education Coordination Council. "Military Education Coordination Council Working Group." Presented to House Armed Services Subcommittee on Oversight and Investigations, 17 September 2008.

"The Navy PME Continuum," Navy Briefing and JCS/DJ-7 Briefing, undated. National Defense University. "House Armed Services Committee Oversight and Investigations Subcommittee Review of Professional Military Education." Presented to House Armed Services Subcommittee on Oversight and Investigations, 29 May 2009.

Retherford, Sam. "Professional Military Education Role of OSD." Presented to House Armed Services Subcommittee on Oversight and Investigations, 20 March 2009.

Roesner, Jack. "OPMEP (.01D) Revision Update." Presented to House Armed Services Subcommittee on Oversight and Investigations, 12 February 2009.

U.S. Air Force. "Do We Have Enough Warrior-Scholars? Air Force General Officers with Ph.D.s." Presented to House Armed Services Subcommittee on Oversight and Investigations, 12 May 2009.

________. "Briefing to HASC O & I." Presented to House Armed Services Subcommittee on Oversight and Investigations, 23 March 2009.

U.S. Army Command and General Staff College. "Preparing Leaders for the Future." Presented to House Armed Services Subcommittee on Oversight and Investigations Staff, Ft. Leavenworth, KS, 26-29 May 2009.

U.S. Army Training and Doctrine Command. "TRADOC Command Overview Brief." Presented to House Armed Services Subcommittee on Oversight and Investigations, 30 April 2009.

U.S. Army War College. "Briefing to House Armed Services Subcommittee on Oversight and Investigations on Professional Military Education." Presented to House Armed Services Committee Staff, 6 April 2009.

U.S. Joint Forces Command. "United States Joint Forces Command Overview." Presented to House Armed Services Subcommittee on Oversight and Investigations Staff, Norfolk, VA, 1 May 2009.

U.S. Joint Forces Staff College. "HASC STAFFDEL Visit." Presented to House Armed Services Subcommittee on Oversight and Investigations, 1 May 2009.

U.S. Transportation Command. "Joint PME Review." Presented to House Armed Services Subcommittee on Oversight and Investigations, Scott Air Force Base, Il, 27-28 May 2009.

Wisecup, RADM James. "Command and PME Brief Prepared for House Armed Services Professional Staffers." Presented to House Armed Services Subcommittee on Oversight and Investigations, 7 April 2009.

End Notes

[1] U.S. Congress, House of Representatives, Committee on Armed Services, *Report of the Panel on Military Education of the One Hundredth Congress*, 101st Cong., 1st sess., 1989, No. 4, (The Skelton Report).

[2] *Goldwater-Nichols Department of Defense Reorganization Act of 1986*, P.L. 99-433.

[3] Judith Hicks Stiehm, *The U.S. Army War College: Military Education in a Democracy* (Temple University Press: Philadelphia, PA, 2002), 11. Admiral Chester Nimitz in a speech to students at the Naval War College in 1960. See also Michael Vlahos, *The Blue Sword: The Naval War College and the American Mission*, (The Naval War College Press: Newport, R.I., 1980).

[4] Churchill as cited in U.S. Congress, House of Representatives, Committee on Armed Services, *Report of the Panel on Military Education of the One Hundredth Congress*, 101st Cong., 1st sess., 1989, No. 4., (*The Skelton Report*), 12.

[5] *National Defense Authorization Act for Fiscal Year 1994*, P.L. 103-160, Sec. 921, 30 November 1993.

[6] This list does not include other joint courses for enlisted members or those for even more senior flag officers, nor does it include other service non-resident officer courses.

[7] *The Skelton Report*.

[8] The Subcommittee's study was not the only one addressing PME to be commissioned in 2009. Last year, the Secretary of Defense asked the Defense Science Board to review PME. That effort has not yet begun. (Defense Science Board PME Study Terms of Reference, 16 January 2009). Two broader, but related, outside studies addressed joint officer management policy and the competencies required for the future officer corps. Both the Center for Strategic and International Studies (CSIS) and Center for a New American Security (CNAS) studies were funded by the Smith- Richardson Foundation established in 1935 to support a wide range of projects that inform important public policy debates http://www.srf.org/. These were led by Dr. Maren Leed and Dr. John Nagl, respectively. HASC O&I Meeting, CSIS, 13 May 2009. CSIS released *The Ingenuity Gap: Officer Management for the 21st Century* in January 2010. The CNAS released *Keeping the Edge: Revitalizing America's Military Officer Corps* in February 2010.

[9] Ike Skelton, *Whispers of Warriors: Essays on the New Joint Era*, National Defense University Press, Washington, D.C., 2004, 49.

[10] This background is borrowed heavily from Dr. David A. Tretler, "Background and Basics: U.S. Professional Military Education," Member Briefing, 24 March 2009; Dr. John W. Yaeger's "Developing National Security Professionals," *Joint Forces Quarterly*, Issue 49, 2nd Quarter 2008, 114-120; Charles A. Henning's, "Legislative History of Joint Professional Military Education," Congressional Research Service, 13 March 2009; and Cynthia Watson, *Military Education: A Reference Handbook*, (Westport, CT: Praeger Security International, 2007).

[11] John Noonan, "The Thayer System: After two centuries of success, it might be time to make some changes," *The Weekly Standard (on line edition)*, 1 August 2007. Sylvanus Thayer, the superintendent of HWest PointH from 1817 to 1833, was not the U.S. Military Academy's first chief, but he so thoroughly reorganized it that he is known as the "Father of the Military Academy."

[12] The staff and students at the Naval War College were renowned for their studies and war gaming before World War II which successfully predicted all of the major Pacific events except the use of *kamikaze* pilots,

according to Admiral Chester Nimitz in a speech to students at the Naval War College in 1960. See also, Michael Vlahos, *The Blue Sword: The Naval War College and the American Mission*, (Newport, R.I.: Naval War College Press, 1980). See also, Judith Hicks Stiehm, *The U.S. Army War College: Military Education in a Democracy*, (Philadelphia, PA: Temple University Press, 2002).

[13] Richard Cheney and Bill Taylor, *Professional Military Education: An Asset for Peace and Progress*, A Report of the Center for Strategic and International Studies Study Group on Professional Military Education, Washington D.C., March 1997, 51.

[14] *Goldwater-Nichols Department of Defense Reorganization Act of 1986*, P.L. 99-433, Sec. 401., 100 Stat. 992, 1 October 1986.

[15] *The Skelton Report*, v.

[16] *Ibid.*, 9-10.

[17] Ibid., 7-8.

[18] Ike Skelton, "Joint Professional Education: Are We There Yet?" *Whispers of Warriors*, 3. See also *The Skelton Report*, 55.

[19] *The Skelton Report*, 4 and 127.

[20] *Ibid.*, 58-59 and 105.

[21] Ike Skelton, "Joint Professional Education: Are We There Yet?" *Whispers of Warriors*, 5. See also *The Skelton Report*, 64, and 127-128. Chairman of the Joint Chiefs of Staff, *Officer Professional Military Education Policy (OPMEP)*, Chairman of the Joint Chiefs of Staff Instruction (CJCSI) 1800.01D, 15 July 2009. The current OPMEP only requires one officer from each non-host service be assigned to each seminar (B-2).

[22] *The Skelton Report*, 4, 64, and 127.

[23] This is a CJCS policy (with "OSD Concurrence for Publication on 7 June 1990") rather than a statutory designation. The earliest reference the Subcommittee found to "single-phase JPME" was in the Office of the Chairman of the Joint Chiefs of Staff CM 344-90, *Military Education Policy Document*, 1 May 1990, II-3, II-11, and III-1 and 2.

[24] *National Defense Authorization Act for Fiscal Year 2005*, P.L. 108-375, Sec 532, 28 October 2004. Section 532 precludes JPME II by being offered by distance learning but is silent on JPME I via this method. This treatment inherently gave the Department the flexibility to deliver JPME I via distance.

[25] *The Skelton Report*, 64 and 67. *OPMEP*, B-1 and B-2.

[26] Ike Skelton, "Joint Professional Education: Are We There Yet?" *Whispers of Warriors*, 8-9.

[27] *OPMEP*.

[28] *OPMEP*, B-5 – B-7 and A-A-5.

[29] *OPMEP*. Also, Chief, Joint Education & Doctrine Division, JCS/DJ7, HASC O&I Interview, 1 October 2009. Joint Advanced Warfighting School (JAWS) used to award credit for both intermediate- and senior-level PME. As of the new *OPMEP, 1800.01D*, JAWS will only offer SLE.

[30] Government Accountability Office (GAO), *Department of Defense: Professional Military Education at Three Senior Service Schools*, GAO/NSIAD-91-202, June 1991. GAO, *Department of Defense: Professional Military Education at the Four Intermediate Service Schools*, GAO/NSIAD-91-182, June 1991. GAO, *Air Force: Status of Recommendations on Officers' Professional Military Education*, GAO/NSIAD-91-122BR, March 1991. GAO, *Navy: Status of Recommendations on Officers' Professional Military Education*, GAO/NSIAD-91-124BR, March 1991. GAO, *Army: Status of Recommendations on Officers' Professional Military Education*, GAO/NSIAD-91-121BR, March 1991. GAO, *Marine Corps: Status of Recommendations on Officers' Professional Military Education*, GAO/NSIAD-91-88FS, February 1991. GAO, *Military Education: Curriculum Changes at the Armed Forces Staff College*, GAO/NSIAD-91-288, September 1991. GAO, *Military Education: Implementation of Recommendations at the Armed Forces Staff College*, GAO/NSIAD-92-30, October 1991. GAO, *Military Education: Implementation of Recommendations at the National War College*, GAO/NSIAD-92-202, June 1992. GAO, *Military Education: Implementation of Recommendations at the Industrial College of the Armed Forces*, GAO/NSIAD-92-221, June 1992. GAO, *Military Education: Actions on Recommendations Involving Institute for Strategic Studies and Capstone*, GAO/NSIAD-92-280, September 1992. GAO, *Military Education: DOD Needs to Enhance Performance Goals and Measures to Improve Oversight of Military Academies*, GAO-03- 1000, September 1993.

[31] Dr. Andrew F. Krepinevich, Written Testimony, Congress, U.S. House of Representatives, Committee on Armed Services, Hearing before the Subcommittee on Oversight and Investigations, *Afghanistan and Iraq: Perspectives on U.S. Strategy, Part III*, 17 November 2009, 8.

[32] See e.g., *2010 Quadrennial Defense Review Report*, February 2010; *National Defense Strategy*, June 2008; *The National Security Strategy of the United States of America*, March 2006; and *The National Military Strategy of the United States of America*, 2004.

[33] U.S. Department of Defense, "Terms of Reference Fact Sheet," *2010 Quadrennial Defense Review*, 27 April 2009, 1; see also *2010 Quadrennial Defense Review Report*, February 2010, 6-9.

[34] *The Skelton Report*, 7.

[35] See *2010 Quadrennial Defense Review Report*, 5.

[36] *Department of Defense Strategic Plan for Joint Officer Management and Joint Professional Military Education*, 3 April 2006, 11, note 24.
[37] *Ibid.*
[38] See Dr. John A. Nagl and Brian M. Burton "The Future of the U.S. Military Officer Corps: Strategic Context," *Keeping the Edge: Revitalizing America's Military Officer Corps*, February 2010; see also Frank G. Hoffman, "Hybrid Warfare and Challenges," *Joint Forces Quarterly*, Issue 52, 1st Quarter, 2009.
[39] Chairman of the Joint Chiefs of Staff, *Capstone Concept for Joint Operations* (CCJO), Version 3.0, 15 January 2009, 4.
[40] *Ibid.*, 4.
[41] *Ibid.*, 12-21 and 28-35.
[42] *Ibid.*, 21.
[43] *Ibid.*, 16.
[44] *Ibid.*, 16-18.
[45] *Ibid.*, 18-20, and note 19.
[46] *Ibid.*, 20.
[47] *Ibid.*, 21.
[48] *Ibid.*, 12-21.
[49] *Ibid.*, 32.
[50] *Ibid.*
[51] Chairman of the Joint Chiefs of Staff, *Capstone Concept for Joint Operations* (CCJO), Version 3.0, 15 January 2009, 2.
[52] *Ibid.*, 5.
[53] Department of Defense, *National Defense Strategy*, June 2008, 22.
[54] *The Joint Operating Environment (JOE) 2010*, 12-37.
[55] *Ibid.*, 38-59.
[56] *Ibid.*, 60-70.
[57] Remarks by General James Mattis, USMC, to the Center for a New American Security, 18 February 2010, 3.
[58] *Ibid.*, 5.
[59] *The Joint Operating Environment (JOE) 2010*, 69.
[60] *Ibid.*, 70.
[61] *Capstone Concept for Joint Operations*, Version 3.0, 34.
[62] *The Joint Operating Environment (JOE) 2010*, 70.
[63] *CJCS Vision for Joint Officer Development*, November 2005, 3.
[64] *Joint Officer Management and Joint Professional Military Education*, 11. See also *CJCS Vision for Joint Officer Development*, November 2005, 4.
[65] See 10 U.S.C. §§3013, 3032, 3033, 5013, 5032, 5033, 5042, 8013, 8032, and 8033.
[66] 10 U.S.C. §153(a)(5)(B) and (C), and *OPMEP*, 15 July 2009, 2.
[67] *CJCS Vision for Joint Officer Development*, November 2005, 5.
[68] *Ibid.*, 3.
[69] *Ibid.*, 9.
[70] See *Chairman OPMEP*, A-A-6.
[71] *CJCS Vision for Joint Officer Development*, 9 (emphasis in the original).
[72] *Ibid.*, 9, and *OPMEP*, 1 and A-1 - A-2.
[73] *OPMEP*, A-1.
[74] *CJCS Vision for Joint Officer Development*, November 2005, 9.
[75] See *OPMEP*, Enclosure E.
[76] Lieutenant General William B. Caldwell IV, Response to Questions for the Record, Congress, U.S. House of Representatives, Committee on Armed Services, Hearing before the Subcommittee on Oversight and Investigations, *Investing in Our Military Leaders: The Role of Professional Military Education in Officer Development*, 28 July 2009, CHARRTS No. HASCOI-07-009.
[77] Caldwell, HASC O&I Hearing, *Investing in Our Military Leaders*, Written Testimony, 3.
[78] Scott Lutterloh, HASC O&I Hearing, *Investing in Our Military Leaders*, Written Testimony, 11, and Response to Question for the Record, CHARRTS No. HASCOI-07-019.
[79] Major General (select) Melvin Spiese, HASC O&I Hearing, *Investing in Our Military Leaders*, Response to Question for the Record, CHARRTS No. HASCOI-07-028.
[80] Spiese, Written Testimony, HASC O&I Hearing, *Investing in Our Military Leaders*, 16.
[81] Dan Sitterly, HASC O&I Hearing, *Investing in Our Military Leaders* Response to Question for the Record, CHARRTS No. HASCOI-07-036.
[82] See 10 U.S.C. §2152.
[83] *OPMEP*, C-1.
[84] *Ibid.*
[85] *Ibid.*, C-2.

[86] *Ibid.*, C-3.
[87] CJCS Memorandum to the Service Chiefs and NDU-P on 2009 Joint Professional Military Education (JPME) Special Areas of Emphasis (SAEs), (CM-0697-09), 23 April 2009, Enclosures 1-16.
[88] See 10 U.S.C. §153; 10 U.S.C. §§2151-2157; and, *OPMEP*, C-4.
[89] 10 U.S.C. §661(a).
[90] 10 U.S.C. §668(a)(1).
[91] See 10 U.S.C. §668(a)(2).
[92] 10 U.S.C. §661(a).
[93] See *Department of Defense Instruction Number 1300.19*, "DoD Joint Officer Management Program," 31 October 2007 (incorporating changes through 16 February 2010). See also *Chairman of the Joint Chiefs of Staff Instruction 1300.05*, "Joint Officer Management Program Procedures," 1 May 2008.
[94] See *Goldwater-Nichols*, Sec. 401 and 10 U.S.C. §668.
[95] 10 U.S.C. §668(b)(1)(A).
[96] 10 U.S.C. §661(d)(2).
[97] See 10 U.S.C. §661(d).
[98] See *National Defense Authorization Act for Fiscal Years 1988 and 1989*, P.L. 100-180, Sec. 1302(a)(1), 4 December 1987 (amending 10 U.S.C. §661(d)(1)). See also *Goldwater-Nichols*, Sec. 401; *The Skelton Report*, 4; and, *National Defense Authorization Act for Fiscal Year 1996*, P.L. 104-106, Sec. 501(a), 10 February 1996 (amending 10 U.S.C. §661 to substitute 800 for the original 1,000 critical positions required by Goldwater-Nichols Sec. 401(d)).
[99] 10 U.S.C. §661(d)(1).
[100] Booz Allen Hamilton, *Independent Study of Joint Officer Management and Joint Professional Military Education*, 17 March 2003, 29-30.
[101] *Ibid.*
[102] *The Skelton Report*, 17.
[103] *Independent Study of Joint Officer Management and Joint Professional Military Education*, 30.
[104] See *Chairman of the Joint Chiefs of Staff Instruction (CJCSI) 1330.05*, "Joint Officer Management Program Procedures," 1 May 2008, Enclosure D.
[105] Joint Staff Legislative Liaison, Written Response to HASC O&I Subcommittee Request for Information, 25 February 2010, Attachment 7, "2008 JDAL Positions by Paygrade," 1-2.
[106] Joint Staff, Briefing to HASC Staff, 6 November 2009, and *Department of Defense Instruction Number 1300.19*, "DoD Joint Officer Management," 31 October 2007 (incorporating changes through 16 February 2010), 14-15 and 24-25.
[107] Joint Staff, Briefing to HASC Staff, 6 November 2009.
[108] Joint Staff Legislative Liaison, Written Response to HASC O&I Subcommittee Request for Information, 25 February 2010, Attachment 1, "JDAL Positions by Category," 1.
[109] *Ibid.*, Attachment 2, "JDAL Positions by Organizations and Service," 1-2.
[110] *Ibid.*, Attachment 7, "2008 JDAL Positions by Paygrade," 1-2.
[111] *National Defense Authorization Act for Fiscal Year 2005 House Report 108-491*, 318.
[112] *Ibid.*, 318; see also *National Defense Authorization Act for Fiscal Year 2005 Conference Report 108-767*, 671.
[113] *Department of Defense Strategic Plan for Joint Officer Management and Joint Professional Military Education*, 12-13.
[114] *Ibid.*, 17 and 36.
[115] *National Defense Authorization Act for Fiscal Year 2007 Conference Report 109-702*, (amending 10 U.S.C. §661), 106-108.
[116] *Report to Congress on Joint Officer Management*, 30 April 2009, 3.
[117] *Goldwater-Nichols*, Sec. 401.
[118] *Department of Defense Strategic Plan for Joint Officer Management and Joint Professional Military Education*, 16.
[119] Joint Staff, Briefing to HASC Staff, 6 November 2009; and CJCSI *1330.05*, "Joint Officer Management Program Procedures," 1 May 2008, Enclosures E (E-1 – E-6) and H (H-1 – H-5). See also *Department of Defense Instruction Number* (DODI) *1300.19*, "DoD Joint Officer Management," 31 October 2007 (incorporating changes through 16 February 2010), Enclosures 5, 6, and 7, 26-37.
[120] Joint Staff, Briefing to HASC Staff, "Joint Officer Management, Joint Duty Assignment List, and Joint Qualification System," 6 November 2009, slide 12.
[121] 10 U.S.C. §661(b)(1)(A).
[122] *DODI 1300.19*, 16.
[123] See *Ibid.*, 16. According to the OPMEP, the joint material at that level should include (1) national military capabilities and organization, the foundation of joint warfare, joint warfare fundamentals, and joint campaigning, E-B-1 – E-B-3.
[124] *DODI 1300.19*, 16.
[125] *Ibid.*, 17.

[126] Joint Staff, Briefing to HASC Staff, 6 November 2009.
[127] *National Defense Authorization Act for Fiscal Year 2007*, P.L. 109-364, Sec. 516, 17 October 2006 (codified as 10 U.S.C. §661(d)).
[128] Joint Staff Legislative Liaison, Joint Duty Assignment List Briefing, 6 November 2009.
[129] *Department of Defense Strategic Plan for Joint Officer Management and Joint Professional Military Education*, 11 April 2006, 11 (citing the *CJCS Vision for Joint Officer Development*, November 2005).
[130] *The Skelton Report*, 11-12.
[131] See *Goldwater-Nichols*, Sec. 401.
[132] See *Goldwater-Nichols*, Sec. 401. See also *National Defense Authorization Act for Fiscal Years 1988 and 1989*, P.L. 100-180, Sec. 1301(b), Sec. 1302(a), 4 December 1987, and *National Defense Authorization Act for Fiscal Year 1989*, P.L. 100-456, Sec. 511, Sec. 512(a), 29 September 1988. Congress also granted the Department authority to issue waivers and make exceptions for JPME completion as well as authority to designate a limited number of officers as JSOs even if they completed their JPME and joint duty assignments out of sequence.
[133] *The Skelton Report*, 52.
[134] *Ibid.*, 102.
[135] *Goldwater-Nichols*, Sec. 401.
[136] *Ibid.*
[137] *National Defense Authorization Act for Fiscal Year 1994, Conference Report 103-357*, Sec. 933, 195; see also 10 U.S.C. §663(b)(2).
[138] See 10 U.S.C. §663.
[139] *National Defense Authorization Act for Fiscal Year 2007, Senate Report 109-254*, 317.
[140] *The Skelton Report*, 57 (emphasis in the original).
[141] *Ibid.* (emphasis added).
[142] *Ibid.*, 58.
[143] *Ibid.*
[144] See 10 U.S.C. §§2152(c) and 2155(b). See also *National Defense Authorization Act for Fiscal Years 1990 and 1991*, Sec. 1123.
[145] 10 U.S.C. §2152(c) (originally 10 U.S.C. §663(c) pursuant to *Goldwater-Nichols*, Sec. 401).
[146] *National Defense Authorization Act for Fiscal Years 1990 and 1991*, P.L. 101-189, Sec. 1123, 29 November 1989.
[147] *National Defense Authorization Act for Fiscal Year 2005, House Report 108-491*, 319-320.
[148] 10 U.S.C. §2155(b).
[149] See 10 U.S.C. §§661-668.
[150] See 10 U.S.C. §661, and *National Defense Authorization Act for Fiscal Year 2007, Conference Report 109-702*, 107-111 and 711-712.
[151] See 10 U.S.C. §§661-663 and §668. See also *National Defense Authorization Act for Fiscal Year 2007, Conference Report 109- 702*, 107-111 and 711-712.
[152] *The Skelton Report*, 16.
[153] See *National Defense Authorization Act for Fiscal Years 1988 and 1989*, Sec. 1302(a)(1), (amending 10 U.S.C. §661(d)(1)). See also *Goldwater-Nichols*, Sec. 401, and *The Skelton Report*, 4.
[154] *The Skelton Report*, 17, and Joint Staff Legislative Liaison, Written Response to HASC O&I Subcommittee Request for Information, 27 February 2010, Attachment 7, "2008 JDAL Positions by Paygrade," 1-2.
[155] 10 U.S.C. §661(d)(1). See also 10 U.S.C. §661(b)-(c).
[156] Joint Staff, Briefing to HASC Staff, 6 November 2009.
[157] *Ibid.*
[158] See 10 U.S.C. §619a.
[159] See *National Defense Authorization Act for Fiscal Year 2005, Conference Report 108-767*, 91-92.
[160] Joint Staff DJ-7/Joint Training Division, "The Joint Staff Officer Project" Final Report, April 2008, 5; HASC O&I Subcommittee Staff Visit to U.S. Southern Command, 6 May 2009; HASC O&I Subcommittee Staff Visit to U.S. Transportation Command and U.S. Strategic Command, 27-28 May 2009; and, HASC Staff Visit to U.S. Northern Command, 30 June 2009.
[161] "The Joint Staff Officer Project," 5.
[162] *Ibid.*, 15.
[163] *Ibid.*, 10 (emphasis added).
[164] See 10 U.S.C. §§661-668.
[165] Joint Staff Legislative Liaison, Written Response to HASC O&I Subcommittee Request for Information, 25 February 2010, Attachment 7, "2008 JDAL Positions by Paygrade," 1-2.
[166] *Ibid.*, 1. See also "The Joint Staff Officer Project," 22-23, and *Department of Defense Instruction Number 1300.19*, "DoD Joint Officer Management," 31 October 2007 (incorporating changes through 16 February 2010).
[167] Joint Staff DJ-7, Briefing to HASC Staff on "The Joint Staff Officer Project," 24 September 2009, slide 4.

[168] "The Joint Staff Officer Project," 10 (emphasis in the original).
[169] *Ibid.*, 57.
[170] See *Joint Staff Information Paper*, 21 September 2009.
[171] See "The Joint Staff Officer Project," 6, 57, and 59-61. (The Fenty Study reported that "a very small percentage of those arriving for duty who were eligible to attend JPME II had actually attended." Only approximately 19 percent of the O-4s surveyed by the study had completed it, but almost 23 percent of all participants had attended JPME II. In addition report, "Officers and senior leaders also felt JPME II (or some version of it) should be part of mandatory preparation—regardless of the rank of incoming personnel—immediately prior to or en route to a Combatant Command assignment.")
[172] "The Joint Staff Officer Project," 57.
[173] *Ibid.*, 58.
[174] See *Ibid.*
[175] See *Ibid.*, 23.
[176] See 10 U.S.C. §661(d)(2).
[177] See *Goldwater-Nichols*, Sec. 401.
[178] 10 U.S.C. §2155(b)(2)(A).
[179] 10 U.S.C. §2152(c).
[180] Joint Staff, Briefing to HASC Staff, 6 November 2009. See also the "Joint Staff Officer Project," 59.
[181] Joint Staff Legislative Liaison, Written Response to HASC O&I Subcommittee Request for Information, 25 February 2010, Attachment 1, "2008 JDAL Positions by Category," 1.
[182] Joint Staff, Briefing to HASC Staff, 17 September 2009.
[183] *Ibid.*
[184] "The Joint Staff Officer Project," 7.
[185] *Ibid.*, 76.
[186] *Ibid.*
[187] *Ibid.*, 37.
[188] "Joint Staff Officer Project," 80.
[189] *Ibid.*, 80.
[190] *Ibid.*, 84.
[191] *Ibid.*, 85.
[192] *Ibid.*, 86.
[193] *Ibid.*, 87.
[194] *Ibid.*, 91.
[195] "The Joint Staff Officer Project," 16-17, 19-20, and 58.
[196] *Ibid.*, 85-87.
[197] David M. Rodney, Christine H. Fox, Samuel D. Kleinman, Michael J. Moskowitz, and Mary E. Lauer, "Developing an Education Strategy for URL Officers," Center for Naval Analysis, March 2008, 1.
[198] *Ibid.*, 1.
[199] *Ibid.*
[200] *Ibid.* Judith Hicks Stiehm also discusses Army War College graduate and supervisors surveys in her 2002 *The U.S. Army War College: Military Education in a Democracy* (Temple University Press: Philadelphia, PA), 79-81, finding similarly troubling results.
[201] "The Joint Staff Officer Project," 59.
[202] Dan Sitterly, Response to Question for the Record, U.S. Congress, House of Representatives, Committee on Armed Services, Subcommittee on Oversight & Investigations Hearing, *Investing in Our Military Leaders: The Role of Professional Military Education in Officer Development*, 28 July 2009, CHARRTS No.: HASCOI-07-038; MajGen (sel) Spiese, Response to Question for the Record, CHARRTS No.: HASCOI-07-029; and LTG William B. Caldwell IV, Response to Question for the Record, CHARRTS No.: HASCOI-07-010.
[203] Scott Lutterloh, Response to Question for the Record, HASC O&I Subcommittee Hearing, *Investing in Our Military Leaders*, CHARRTS No.: HASCOI-07-020.
[204] *Ibid.*
[205] Memorandum for NORAD and USNORTHCOM Directors, Policy Memorandum, "Joint Professional Military Education (JPME) Phase II," August 27, 2008.
[206] *Ibid.*
[207] U.S Central Command, "Command Policy Letter Number 40," 9 April 2009.
[208] HASC O&I Subcommittee Staff Visit to U.S. Central Command, 4-7 May 2009.
[209] HASC O&I Subcommittee Staff Visit to U.S. Southern Command, 4-7 May 2009.
[210] *CJCS Vision for Joint Officer Development*, November 2005, 6. See also 10 U.S.C. §2154.
[211] Department of Defense Legislative Proposal to the HASC (draft), submitted April 2010.
[212] HASC O&I Subcommittee Staff Visit to Joint Forces Staff College, 1 May 2009.

[213] Gordon Nathaniel Lederman, *Reorganizing the Joint Chiefs of Staff: The Goldwater-Nichols Act of 1986*, (Westport, CT: Greenwood Press, 1999), 107.
[214] LTG Caldwell, Written Testimony, HASC O&I Subcommittee Hearing, *Investing in Our Military Leaders*.
[215] General Martin Dempsey, USA, Commanding General, Army Training and Doctrine Command, Meeting with HASC O&I Subcommittee Staff, 30 April 2009.
[216] Paxton, HASC O&I Subcommittee Hearing, *Investing in Our Military Leaders*, Response to Questions for the Record, CHARRTS No: HASCOI-07-001.
[217] Chairman of the Joint Chiefs of Staff Instruction 1300.05, *Joint Officer Management Program Procedures*, 1 May 2008.
[218] Paxton, HASC O&I Subcommittee Hearing, *Investing in Our Military Leaders*, Response to Questions for the Record, CHARRTS No: HASCOI-07-001. An explanation of the "intensity factor" multiplier in the point accrual formula for joint experience points is contained in DODI 1300.19, *DOD Joint Officer Management Program*, 21 August 2008, 18.
[219] *Department of Defense Strategic Plan for Joint Officer Management and Joint Professional Military Education*, 3 April 2006, 13 and 18.
[220] U.S. Joint Forces Command, *The 2010 Joint Operating Environment (JOE)*, 5.
[221] RADM Daniel Davenport and RDML John Richardson, U.S. Joint Forces Command, Briefing to HASC Staff, 20 March 2009, and *The 2010 Joint Operating Environment*.
[222] *The 2010 Joint Operating Environment*, 4.
[223] *Capstone Concept for Joint Operations* (Version 3.0), 28.
[224] *Ibid*., 28 and 30-35.
[225] "Service Responsibilities Regarding Joint Officer Development," *CJCS Vision for Joint Officer Development*, November 2005, 2.
[226] *Ibid*., 8 (emphasis in the original).
[227] *Capstone Concept for Joint Operations* (Version 3.0), 28 (emphasis added).
[228] U.S. Congress, House of Representatives, Committee on Armed Services, Subcommittee on Oversight & Investigations, *Agency Stovepipes vs Strategic Agility: Lessons We Need to Learn from Provincial Reconstruction Teams in Iraq and Afghanistan*, April 2008, No. 8, and *Stand Up and Be Counted: The Continuing Challenge of Building the Iraqi Security Forces*, June 2007, 131-133.
[229] *OPMEP*, E-B-1 through 3.
[230] *Ibid*., E-C-1 through 4.
[231] *OPMEP* (22 December 2005), E-B-3.
[232] Joint Doctrine and Education Division, Operational Plans and Joint Force Development Directorate (DJ-7), HASC O&I Subcommittee Staff Briefing, 1 October 2009.
[233] Student Panel, HASC Staff Visit, College of Naval Command and Staff, 7 April 2009.
[234] *The Skelton Report*, 35.
[235] *Ibid*., 33.
[236] *Ibid*., 33-34.
[237] 10 U.S.C. §2155 (c)(1).
[238] *OPMEP*, E-E-1.
[239] 10 U.S.C. §2155 (c)(1), 10 U.S.C. §2151 (a)(1).
[240] *The Skelton Report, 30*-32.
[241] *Ibid*., 31.
[242] General George Casey, Chief of Staff, U.S. Army, HASC O&I Subcommittee Members Meeting, 21 October 2009.
[243] *Ibid*. See also, Army OCLL email to HASC staff, 8 April 2010.
[244] David H. Petraeus, "Beyond the Cloister," The National Interest, July/August 2007. Accessed online at http://www.the-american-interest.com/ai2/article.cfm?Id=290&MId=1 (12 November 2009), 2.
[245] *Ibid*.
[246] Lieutenant General (Ret.) David Barno, Written Testimony, Congress, U.S. House of Representatives, Committee on Armed Services, Hearing before the Subcommittee on Oversight and Investigations, *Charting the Course for Effective Professional Military Education*, 10 September 2009.
[247] Professor Williamson Murray, Testimony, HASC O&I Subcommittee Hearing, *Charting the Course*.
[248] *The Skelton Report*, 31.
[249] Congress, U.S. House of Representatives, Committee on Armed Services, Hearing before the Subcommittee on Oversight and Investigations, *Another Crossroads? Professional Military Education 20 Years After the Goldwater-Nichols Act and the Skelton Panel*, 20 May 2009.
[250] Dr. Janet Breslin-Smith, Testimony, HASC O&I Subcommittee Hearing, *Another Crossroads?*
[251] Dr. John Williams, Testimony, HASC O&I Subcommittee Hearing, *Charting the Course*.
[252] General James Conway, Commandant, U.S. Marine Corps, HASC O&I Subcommittee Members Meeting, 24 September 2009.
[253] HASC O&I Subcommittee Hearing, *Charting the Course*.

[254] HASC O&I Members Meeting, Admiral Michael Mullen, 27 October 2009.
[255] Murray, Testimony, HASC O&I Subcommittee Hearing, *Charting the Course*.
[256] 10 U.S.C. §§2161, 2163, 4314, 4321, 7048, 7101, 7102, 9314, and 9317.
[257] Dan Sitterly, Testimony, HASC O&I Subcommittee Hearing, *Investing in Our Military Leaders*.
[258] Dr. Stephen Chiabotti, Dean, School of Advanced Air and Space Studies, Telephone Interview, 18 August 2009.
[259] Col. Mace Carpenter and Dr. Sorin Lungu, HASC O&I Subcommittee Staff Briefing, "School of Advanced Strategic Planning and Strategy (ASPS)," 12 August 2009.
[260] Colin S. Gray, "On Strategic Performance," *Joint Forces Quarterly*, Winter 1995-1996, 34.
[261] *The Skelton Report*, 28.
[262] *Ibid.*, 31.
[263] *Ibid.*, 39.
[264] *Ibid.*, 29 and 39-41.
[265] *Ibid.*, 40.
[266] *Ibid.*
[267] *Ibid.*
[268] See Arthur F. Lykke, Jr., "Chapter 13: Toward and Understanding of Military Strategy," *U.S. Army War College Guide to Strategy*, February 2001, 179. See also Carl von Clausewitz, *On War*, trans. and ed. by Michael Howard and Peter Paret, (Princeton, NJ: Princeton University Press, 1976), 69, 75, and 177 (defining strategy as "the use of the engagement for the purpose of the war," wherein Clausewitz recognized the term "war" both as "an act of force to compel our enemy to do our will," and "nothing but the continuation of policy with other means."); Julian S. Corbett, *Some Principles of Maritime Strategy*, (New York, NY: Longmans, Green & Co., 1911), 308 ("Strategy is the art of directing force to the ends in view."); B.H. Liddell Hart, *Strategy: The Indirect Approach*, (London: Faber and Faber, 1954), 335 (defining strategy as "the art of distributing and applying military means to fulfil [sic] the ends of policy"); Gabriel Marcella and Stephen O. Fought, "Teaching Strategy in the 21st Century," *Joint Forces Quarterly*, Issue 52, 1st Quarter 2009, 57 ("Strategy is the art of applying power to achieve objectives within the limits imposed by policy"); and, Colin S. Gray, "Schools for Strategy: Teaching Strategy for 21st Century Conflict," *Strategic Studies Institute*, 28 October 2009, 7 ("I choose to define (military) strategy as the direction and use that is made of force and the threat of force for the ends of policy.").
[269] See Michael J. Mazar and Colonel Nelson Cabot, USAF, *Introduction to Strategy, Core Course 6100 Academic Year 2008-2009*, The National War College, 24.
[270] "Strategy," *Joint Publication 1-02: Department of Defense Dictionary of Military and Associated Terms*.
[271] See Bernard Brodie, *War and Politics*, (New York, NY: Macmillan, 1973), 452.
[272] *The Skelton Report*, 25.
[273] See Samuel P. Huntington, *The Soldier and the State: The Theory and Politics of Civil-Military Relations*, (Cambridge, MA: Harvard University Press, 1957), 262-263, 307-309, and 418-423; *The Skelton Report, 23*; *Introduction to Strategy*, The National War College, 14; and, HASC O&I Hearing, *Investing in our Military Leaders*.
[274] Gray, "On Strategic Performance," 34.
[275] *The Skelton Report*, 26.
[276] *Ibid.*
[277] *Ibid.*, (citing the 1982 *Army Field Manual (FM 100-5)* definition of 'Operational Art').
[278] *Ibid.*
[279] *Capstone Concept for Joint Operations*, 34.
[280] U.S. Joint Forces Command, *The Joint Operating Environment (JOE) 2010*, 18 February 2010, 70.
[281] *Ibid.*, 62 and 18.
[282] See Colin S. Gray, *Schools for Teaching Strategy*, 8. See also Marcella and Fought, "Teaching Strategy in the 21st Century," 57.
[283] See U.S. Joint Forces Command, *The Joint Operating Environment (JOE) 2010*, 18 February 2010, 4.
[284] *Ibid.*, 2.
[285] *Ibid.*, 71.
[286] Murray, HASC O&I Subcommittee Hearing, *Charting the Course*, Response to Question for the Record, 1-2.
[287] *The Skelton Report*, 27-28.
[288] *Ibid., 28.*
[289] *Ibid.*
[290] HASC O&I Subcommittee Hearing, *Investing in our Military Leaders*. See also RADM James P. Wisecup, Response to Question for the Record, HASC O&I Subcommittee Hearing, C*harting the Course*, CHARRTS No.:HASCOI-06-008, 3, and Professor Milan Vego, "There's No Place Like Newport," *Proceedings*, February 2010, 40.
[291] See, e.g., HASC O&I Subcommittee Hearings, *Investing in our Military Leaders* and *Charting the Course*.
[292] Colin S. Gray, *Schools for Teaching Strategy: Teaching Strategy for 21st Century Conflict*, November 2009, 31.
[293] *Ibid.*

[294] Joint Staff Legislative Liaison, Written Response to HASC O&I Subcommittee Request for Information, 13 November 2009, Attachment 6.
[295] *OPMEP*, D-1.
[296] *Ibid.*, A-1 (emphasis added).
[297] *Ibid.*, A-A-1 - A-A-2 (emphasis added).
[298] See Gray, *Schools for Teaching Strategy*, v.
[299] *OPMEP*, 2.
[300] HASC O&I Subcommittee Hearing, *Investing in our Military Leaders*.
[301] *Ibid.*
[302] *The Skelton Report*, 39.
[303] *Ibid.*, 39-40.
[304] *Commissioned Officer Professional Development and Career Management*, (DA PAM 600-3), 1 February 2010, 266-267; Army Legislative Liaison, Written Response to a HASC O&I Subcommittee Request for Information, 3 February 2010.
[305] *Commissioned Officer Professional Development and Career Management*, 263.
[306] *Ibid.*, 264.
[307] BG Edward P. Donnelly, HASC O&I Staff Briefing, 7 October 2009; Army Legislative Liaison, Written Response to a HASC O&I Subcommittee Request for Information, 3 February 2010; and, HASC O&I Subcommittee Hearing, *Charting the Course*.
[308] Donnelly, HASC O&I Staff Briefing, 7 October 2009.
[309] *Ibid.*
[310] *Commissioned Officer Professional Development and Career Management*, 266.
[311] *Ibid.*
[312] *Ibid.*, 266-267.
[313] *Ibid.*
[314] *Ibid.*, 266-268.
[315] Donnelly, HASC O&I Staff Briefing, 7 October 2009.
[316] *Ibid. and Commissioned Officer Professional Development and Career Management*, 266, 267, and 269.
[317] *Commissioned Officer Professional Development and Career Management*, 264-265 and 269.
[318] End Notes is not available.
[319] *Ibid.*, 269.
[320] Navy Legislative Liaison, Written Response to a HASC O&I Subcommittee Request for Information, 20 November 2009, 1.
[321] Navy Legislative Liaison, Written Response to a HASC O&I Subcommittee Request for Information, 10 March 2010. Some naval officers hold more than one strategic subspecialty designation.
[322] *Ibid.*
[323] Navy Legislative Liaison, "National Security Studies (2000) Core Skill Requirements," Response to a HASC O&I Subcommittee Request for Information, 25 November 2009, 1.
[324] *Ibid.*,1.
[325] *Ibid.*
[326] HASC O&I Staff Visit to NWC, 7 April 2009.
[327] "National Security Studies (2000) Core Skill Requirements," 1.
[328] RADM Wisecup, Response to Question for the Record, U.S. Congress, House of Representatives, Committee on Armed Services, Subcommittee on Oversight & Investigations Hearing, *Raising Thinking from the Tactical to the Operational Level: JPME I and II at the Services' and Joint Command & Staff Colleges*, 25 June 2009, CHARRTS No.:HASCOI-06-010, 1-2.
[329] RADM Wisecup, Response to Question for the Record, HASC O&I Subcommittee Hearing, *Raising Thinking*, CHARRTS No.:HASCOI-06-008, 3; See also, Professor Milan Vego, "There's No Place Like Newport," 40.
[330] See Admiral James Stavridis and Captain Mark Hagerott, "The Heart of an Officer: Joint Interagency, and International Operations and Navy Career Development," *Naval War College Review*, Volume 62, Number 2, Spring 2009, 30 and 33-34.
[331] Air Force Legislative Liaison, Written Response to a HASC O&I Subcommittee Request for Information, 30 October 2009, 2.
[332] Marine Corps Legislative Liaison, Written response to a HASC O&I Subcommittee Request for Information, 17 November 2009, 4.
[333] *Ibid.*
[334] *The Skelton Report*, 29.
[335] *Ibid.*, 30 (emphasis in the original).
[336] *Ibid.*
[337] *Ibid.*
[338] *Ibid.*, 30.
[339] *Ibid.*, 31.

[340] *Ibid.*, 30.
[341] *Ibid.*, 31.
[342] Army, Navy, Air Force, and Marine Corps Legislative Liaisons, Written Responses to a HASC O&I Subcommittee Requests for Information, 3 November 2009 (USA), 13 and 16 November 2009 (USN), 30 October 2009 (USAF), and 17 November 2009 (USMC).
[343] Barno, Testimony, HASC O&I Subcommittee Hearing, *Charting the Course;* See also, Dr. John A. Nagl and Brian M. Burton "Revitalizing America's Officer Core," *Keeping the Edge: Revitalizing America's Military Officer Corps*, Center for a New American Security, February 2010, 69-70.
[344] *The Skelton Report*, 34-35.
[345] See 10 U.S.C. §2155, and *OPMEP*, A-A-7 - A-A-8 and A-A-A-1.
[346] See 10 U.S.C. §§2151(a) and 2154.
[347] *OPMEP*, A-A-7 and E-C-1 - E-C-3.
[348] See e.g., HASC O&I Subcommittee Hearing, *Investing in Our Military Leaders.*
[349] See 10 U.S.C. §2151(a).
[350] *OPMEP*, A-A-7 and E-C-1 - E-C-3.
[351] *Ibid.*, A-A-7; A-A-A-1; and, E-D-1 - E-D-3.
[352] These programs include the Army's School of Advanced Military Studies (SAMS), the Navy's Maritime Advanced Warfighting School (MAWS), the Air Force's School of Advanced Air and Space Studies (SAASS), and the Marine Corps' School of Advanced Warfighting (SAW). Graduates of MAWS receive a master's degree in national security and strategic studies from the Naval War College; graduates of SAMS earn a master's degree in military arts and sciences; SAASS graduates are awarded a master's degree in airpower arts and sciences from the Air University; and graduates of SAW may elect to earn a master's degree in operational studies.
[353] The Marine Corps limits the focus of SAW to the operational level of warfare.
[354] Barno, Written Testimony, HASC O&I Subcommittee Hearing, *Charting the Course*, 6.
[355] See 10 U.S.C. §2155(c).
[356] *OPMEP*, A-A-7 - A-A-8 and A-A-A-1.
[357] *OPMEP*, A-A-8 and A-A-A-1.
[358] *Ibid*, A-A-7 - A-A-8, and A-A-A-1.
[359] *Ibid.*, A-A-8 and E-H-1 - E-H-3.
[360] *Ibid.*, E-H-1.
[361] RADM Wisecup, *R*esponse to Question for the Record, HASC O&I Subcommittee Hearing, *Raising Thinking*,CHARRTS No.: HASCOI-06-008, 3 (quoting Admiral Stansfield Turner).
[362] HASC O&I Subcommittee Hearing, *Charting the Course.*
[363] See MG Robert Scales, USA (Ret.), "Too Busy to Learn," *Proceedings*, February 2010, 33; Colonel Mace Carpenter, "School of Advanced Strategic Planning and Strategy (ASPS) Proposal," Slide 10.
[364] *OPMEP*, A-A-8; E-K-1 - E-K-2.
[365] See HASC O&I Hearing, *Charting the Course*. See also Paula G. Thornhill, "Flag Officer Education: It's Time to Improve the Capstone Course," *Armed Forces Journal* (online edition), 9 April 2010, 4.
[366] Chairman of the Joint Chiefs of Staff, "CAPSTONE Course Guidance," *Memorandum for the President, National Defense University*, 23 June 2009.
[367] *OPMEP*, E-L-1 and E-L-3
[368] *OPMEP*, E-L-1 - E-L-4.
[369] *Ibid.*, A-A-8 and E-M-1.
[370] General George Casey, Chief of Staff of the U.S. Army, HASC O&I Members Meeting, 21 October 2009; and Admiral Gary Roughead, HASC O&I Members Meeting, Chief of Naval Operations, 29 September 2009.
[371] Barno, Written Testimony, HASC O&I Subcommittee Hearing, *Charting the Course*, 6-7.
[372] *The Skelton Report, 158.*
[373] *Ibid.*
[374] *Ibid.*
[375] *Ibid.*, 158-161.
[376] *OPMEP*, E-1.
[377] The *Skelton Report*, 160.
[378] *OPMEP*, E-1.
[379] DDJS-ME Memorandum, *Process for Accreditation of Joint Education (PAJE), Joint Advanced Warfighting School (JAWS)*, 21-25 April 2008, 4, and DDJS-ME Memorandum, *Process for Accreditation of Joint Education (PAJE), U.S. Air Force Command and Staff College (ACSC)*, 25-29 February 2008, 6.
[380] DDJS-ME Memorandum, *Process for Accreditation of Joint Education (PAJE), U.S. Air Force Command and Staff College (ACSC)*, 25-29 February 2008, 6.
[381] National Leaders, Faculty and Student Panels, HASC Staff Visit, 3 April 2009.

[382] Alan Brinkley, Betty Dessants, Michael Flamm, Cynthia Fleming, Charles Forcey and Eric Rothschild, *The Chicago Handbook for Teachers, A Practical Guide to the College Classroom*, (Chicago, Il: University of Chicago Press, 1999).
[383] CODELs Snyder and HASC Staff Visits to PME Institutions, March – July 2009.
[384] DDJS-ME Memorandum, *Process for Accreditation of Joint Education (PAJE), Marine Corps War College (MCWAR)*, 12-14 January 2009, 5.
[385] DDJS-ME Memorandum, *Process for Accreditation of Joint Education (PAJE), U.S. Air Force Command and Staff College (ACSC)*, 25-29 February 2008, 6.
[386] DDJS-ME Memorandum, *Process for Accreditation of Joint Education (PAJE), College of Naval Warfare (CNW)*, 14-18 May 2007, 6.
[387] DDJS-ME Memorandum, *Process for Accreditation of Joint Education (PAJE), U.S. Army War College (USAWC)*, 21-15 September 2009, 6.
[388] *OPMEP*, B-1 and B-2.
[389] *Ibid.*
[390] DDJS-ME Memorandum, *Process for Accreditation of Joint Education (PAJE), Marine Corps War College (MCWAR)*, 12-14 January 2009, 5.
[391] DDJS-ME Memorandum, *Process for Accreditation of Joint Education (PAJE), College of Naval Warfare (CNW)*, 4-7 May, 2009, 4, and DDJS-ME Memorandum, *Process for Accreditation of Joint Education (PAJE), U.S. Army War College (USAWC)*, 21-25 September 2009, 5.
[392] DDJS-ME Memorandum, *Process for Accreditation of Joint Education (PAJE), College of Naval Command and Staff (CNCS)*, 25-30 October, 2009, 5.
[393] *Skelton Report*, 37.
[394] *Ibid.*, 37.
[395] *Ibid.*, 44.
[396] DDJS-ME Memorandum, *Process or Accreditation of Joint Education (PAJE), College of Naval Command and Staff*, 25-30 October 2009, 1, 5, 11, and 14-16.
[397] General Martin Dempsey, HASC O&I Staff Meeting, 30 April 2009.
[398] *Skelton Report*, 162.
[399] *Ibid.*, 165-166.
[400] *Ibid.*, 162-163.
[401] *Ibid.*, 166.
[402] "USAWC Student Academic Assessment and Evaluation Resident and Distance Education Programs," *Carlisle Barracks Memorandum 623-1*, (Carlisle, PA: USAWC, 11 January 2007), 3.
[403] "Naval War College Standard Organization and Regulations Manual (SORM)," *NWC Instruction 1520.2* (Newport, RI), Annex A-2 - C-3.
[404] *Ibid.*, Annex A-2 - C-5.
[405] CDR Thomas Sliski, NWC, Email to HASC O&I Subcommittee Staff, 26 January 2010.
[406] National War College *Student Handbook 2008-2009*, 30.
[407] ICAF Policy Paper, "Student Evaluations/Grading Policy: Academic Year 2008-2009," undated.
[408] Student Panel, HASC O&I Staff Visit, CGSC, 27 May 2009.
[409] Breslin-Smith, Response to Question for the Record, HASC O&I Subcommittee Hearing, *Another Crossroads?*
[410] Murray, Response to Question for the Record, HASC O&I Subcommittee Hearing, *Charting the Course*.
[411] Williams, Response to Question for the Record, HASC O&I Subcommittee Hearing, *Charting the Course*.
[412] *OPMEP*, E-1.
[413] DDJS-ME Memorandum, *Process for Accreditation of Joint Education (PAJE), USA Command and General Staff College (CGSC)*, 25-29 February 2008, 5.
[414] Student Panel, HASC O&I Staff Visit, CGSC, 27 May 2009.
[415] *Ibid.*
[416] Faculty Panel, HASC O&I Staff Visit, CGSC, 27 May 2009.
[417] Student Panel, HASC Staff Visit, ACSC, 9 April 2009.
[418] Student Panel, HASC O&I Staff Visit, CGSC, 27 May 2009.
[419] Faculty Panel, HASC O&I Staff Visit, CGSC, 27 May 2009.
[420] Major General Robert M. Williams, Congress, U.S. House of Representatives, Committee on Armed Services, Hearing before the Subcommittee on Oversight and Investigations, Response to a Question for the Record, HASC O&I Hearing, *Thinkers and Practitioners: Do Senior Professional Military Education Schools Produce Strategists?* 4 June 2009, CHARRTS No: HASCOI-05-003.
[421] *Ibid.*
[422] *Ibid.*
[423] *Skelton Report*, 79 and 120.
[424] *Ibid.*, 120 and 122.
[425] *OPMEP* (22 December 2005), E-K-1.
[426] *Ibid.*, 4.

[427] Joint Staff DJ-7 Joint Education and Doctrine Division, HASC O&I Staff Briefing, 19 August 2009.
[428] Catapult Consultants, *CAPSTONE Program Independent Review Final Report*, January 2009.
[429] CJCS Memorandum for the President, National Defense University (CM-0764-09), 23 June 2009.
[430] Joint Staff DJ-7 Joint Education and Doctrine Division, HASC O&I Staff Briefing, 19 August 2009.
[431] *Skelton Report*, 162.
[432] *Ibid.*, 68-69.
[433] *Ibid.*, 68.
[434] *OPMEP*, E-1.
[435] Major General Robert Steel, Testimony, HASC O&I Subcommittee Hearing, *Thinkers and Practitioners*.
[436] HASC Staff Visit to Marine Corps University, Quantico, VA, 23 March 2009.
[437] Brigadier General Jimmie Jackson, Written Testimony, Congress, U.S. House of Representatives, Committee on Armed Services, Hearing before the Subcommittee on Oversight and Investigations Hearing, *Raising Thinking from the Tactical to the Operational Level: JPME I and II at the Services' and Joint Command & Staff Colleges*, 25 June 2009.
[438] Faculty Panel, HASC O&I Staff Visit, CGSC, 26 May 2009, and Faculty Panel, HASC O&I Staff Visit, National, 3 April 2009.
[439] Cardon, HASC O&I Subcommittee Hearing, *Raising Thinking*, Written Testimony, 16.
[440] MajGen Steel.Testimony, HASC O&I Subcommittee Hearing, *Thinkers and Practitioners*.
[441] *The Skelton Report*, 24.
[442] RADM Wisecup, Testimony, HASC O&I Subcommittee Hearing, *Thinkers and Practitioners*.
[443] Irregular Warfare (IW): Joint Operating Concept, 11September 2007, 6.
[444] *Ibid.*, 1, and *DOD Dictionary of Military Terms*, Joint Publication 1-02 (current as of 31 October 2009).
[445] Department of Defense, *National Defense Strategy*, June 2008., 4 and 13.
[446] *OPMEP*, A-2 (emphasis added).
[447] *Ibid.*, A-2, 3.
[448] The OPMEP prescribes Joint Learning Areas and Objectives for the levels of service PME and for individual joint schools. Within the Joint Learning Areas and Objectives are numbered Learning Areas for specific requirements.
[449] *Ibid.*, E-C-2.
[450] Brigadier General Edward Cardon, Testimony, HASC O&I Subcommittee Hearing, *Raising Thinking*.
[451] *Ibid.*
[452] Damm, HASC O&I Subcommittee Hearing, *Raising Thinking*, Written Testimony, 14-15.
[453] RADM Wisecup, Testimony, HASC O&I Subcommittee Hearing, *Thinkers and Practitioners*.
[454] Irregular Warfare (IW): Joint Operating Concept, 11September 2007, 10.
[455] *Joint Operations*, Joint Publication 3-0 (with change 1 of 12 February 2008), iv. The six phases are: shape, deter, seize the initiative, dominate, stabilize, and enable civilian authority.
[456] Congress, U.S. House of Representatives, Committee on Armed Services, Hearing before the Subcommittee on Oversight and Investigations, *The Role of the Department of Defense in Provincial Reconstruction Teams (PRTs) in Iraq and Afghanistan*, 5 September 2007.
[457] U.S. Congress, House of Representatives, Committee on Armed Services, Subcommittee on Oversight & Investigations, *Agency Stovepipes vs Strategic Agility: Lessons We Need to Learn from Provincial Reconstruction Teams in Iraq and Afghanistan*, April 2008, 51.
[458] *Department of Defense Directive 3000.05*, "Military Support for Stability, Security, and Transition and Reconstruction," (Washington, DC: 28 November 2005), 2, and *Department of Defense Instruction 3000.05*, "Stability Operations," (Washington, DC: 16 September 2009), 15.
[459] *Quadrennial Defense Review Report*, (Washington, DC: February 2010), viii.
[460] *Ibid.*, 24.
[461] *DODI 3000.05*, "Stability Operations," 2.
[462] *Quadrennial Defense Review Report*, xiii.
[463] *OPMEP*, E-C-3. The OPMEP makes no mention of stability operations but does use the SSTR acronym with encompasses stability operations.
[464] DDJS-ME Memorandum, Process for Accreditation of Joint Education (PAJE), U.S. Army War College, 21-25 September 2009, 12-19.
[465] DDJS-ME Memorandum, Process for Accreditation of Joint Education (PAJE), College of Naval Command and Staff, 25-30 October 2009, 16-17.
[466] *DODD 3000.05*, "Military Support for Stability, Security, and Transition and Reconstruction," 2.
[467] Brigadier General Katherine P. Kasun, Response to Question for the Record, HASC O&I Subcommittee Hearing, *Raising Thinking*, CHARRTS No: HASCOI-06-005.
[468] Brigadier General Edward Cardon, Response to Question for the Record, HASC O&I Subcommittee Hearing, *Raising Thinking*, CHARRTS No: HASCOI-06-005.
[469] *Ibid.*

[470] HASC RADM Wisecup, Response to Questions for the Record, O&I Subcommittee Hearing, *Raising Thinking*, CHARRTS No: HASCOI-06-005.

[471] Colonel Raymond Damm, Response to Question for the Record, HASC O&I Subcommittee Hearing, *Raising Thinking*, CHARRTS No: HASCOI-06-005.

[472] *Language and Regional Expertise Planning*, Chairman, Joint Chiefs of Staff Instruction (CJCSI) 3126.01 (current as of 11 February 2008), A-1.

[473] U.S. Congress, House of Representatives, Committee on Armed Services, Subcommittee on Oversight & Investigations, *Building Language Skills and Cultural Competency in the Military: DOD's Challenge in Today's Educational Environment*, November 2008.

[474] *OPMEP*, Enclosure E.

[475] *Quadrennial Defense Review Report*, 18 and 25.

[476] ACSC Staff, Response to Questions for the Record, HASC O&I Subcommittee Hearing, *Raising Thinking*, CHARRTS No: HASCOI-06-023.

[477] Major General Maurice Forsyth, Testimony, HASC O&I Subcommittee Hearing, *Thinkers and Practitioners*.

[478] ACSC Staff, Response to Questions for the Record, HASC O&I Subcommittee Hearing, *Raising Thinking*, CHARRTS No: HASCOI-06-023, and HASC Staff Visit, AU, 9 April 2009.

[479] CODEL Snyder to MCU, Quantico, VA, 29 May, 2009.

[480] Col Damm, Response to Questions for the Record, HASC O&I Subcommittee Hearing, *Raising Thinking*, CHARRTS No: HASCOI-06-023.

[481] CODEL Snyder to MCU, Quantico, VA, 29 March, 2009.

[482] RADM Wisecup, Testimony, HASC O&I Subcommittee Hearing, *Thinkers and Practitioners*.

[483] O&I Subcommittee staff discussions with Naval War College Staff and Faculty, Newport, RI, 6 April 2009.

[484] BG Cardon, Testimony, HASC O&I Subcommittee Hearing, *Raising Thinking*.

[485] *OPMEP*, E-C-2 and E-D-2.

[486] Rear Admiral Garry Hall, Testimony, HASC O&I Subcommittee Hearing, *Thinkers and Practitioners*.

[487] RADM Wisecup, Testimony, HASC O&I Subcommittee Hearing, *Thinkers and Practitioners*.

[488] RADM Hall, Testimony, HASC O&I Subcommittee Hearing, *Thinkers and Practitioners*.

[489] Colonel Michael Belcher, Testimony, HASC O&I Subcommittee Hearing, *Thinkers and Practitioners*.

[490] "Marine Corps University Catalog," Academic Year 2008-2009, Marine Corps University, Quantico, VA, 42.

[491] BG Cardon, Email to HASC O&I Subcommittee, 19 June 2009.

[492] William Spain, Email to HASC O&I Subcommittee, 19 June 2009.

[493] The definitions of the terms describing the JIIM environment are contained in Joint Publication 3-16, *Joint Doctrine for Multinational Operations*, 7 March 2007 and further developed in "Turning Ideas into Action," *Project on National Security Reform*, September 2009, accessed online on 23 March 2010 at http://www.pnsr.org/data/files/pnsr_turning_ideas_into_action.pdf.

[494] *OMPEP* (1 December 2000), E-B-1, 2, 3.

[495] *OPMEP* (22 December 2005), E-D-2.

[496] *Ibid*., E-C-2.

[497] *Ibid*., E-D-2.

[498] Testimony of Admiral James G. Stavridis, USN, Commander, United States European Command before the House Armed Services Committee, 10 March 2010.

[499] Posture Statement of Admiral James G. Stavridis, USN, Commander, U.S. Southern Command before the House Armed Services Committee, 18 March 2009 and Posture Statement of General Douglas M. Fraser, USAF, Commander, United States Southern Command, 12 March 2010.

[500] Statement of Admiral Robert F. Willard, USN, Commander, United States Pacific Command before the House Armed Services Committee, 25 March 2010 (emphasis added).

[501] Statement of Admiral Robert F. Willard, USN, Commander, United States Pacific Command before the House Armed Services Committee, 25 March 2010.

[502] *Skelton Report*, 30.

[503] *Ibid*., 131.

[504] Major General Robert Williams, Testimony, HASC O&I Subcommittee Hearing, *Thinkers and Practitioners*.

[505] "Marine Corps University Catalog," Academic Year 2008-2009, Marine Corps University, Quantico, VA, 39.

[506] Faculty Panel, HASC O&I Staff Visit, CGSC, 26 May 2009, and Faculty Panel, HASC Staff Visit, 3 April 2009.

[507] Jeffrey D. McCausland and Neil Weissman, *Educating Leaders in an Age of Uncertainty – The Future of Military War Colleges:A Research Study for the Smith Richardson Foundation*, Dickinson College Leadership in Conflict Initiative, Carlisle, Pennsylvania, 15 December 2005, 298-299.

[508] U.S. Department of Defense, "Functions of the Department of Defense and Its Major Components," *Department of Defense Directive 5100.01*, 21 November 2003.

[509] Sam Retherford, Director for Officer and Enlisted Personnel Management, Under Secretary of Defense for Personnel and Readiness, Military Personnel Policy, "Professional Military Education: Role of OSD," Briefing for HASC Staff, 20 March 2009, and Telephone Interview, 9 November 2009. See also Dr. Janet Breslin-Smith, Written Testimony, Congress, U.S. House of Representatives, Committee on Armed Services, Hearing

Before the Subcommittee on Oversight and Investigations, *Another Crossroads? Professional Military Education 20 Years After the Goldwater-Nichols Act and the Skelton Panel*, 20 May 2009, 3. Also, Faculty Panels, CODELs Snyder to ICAF, 31 March 2009, and National, 29 April 2009. Also HASC Staff Visits to National, 3 April 2009, and JFSC, 1 May 2009.

[510] 10 U.S.C. §2152.

[511] Ibid.

[512] 10 U.S.C. §2154.

[513] Ibid.

[514] 10 U.S.C. §2154 and 2155. The Chairman of the Joint Chiefs can grant waivers for the sequencing under "exceptional circumstances" on a case by case basis.

[515] 10 U.S.C. §2155.

[516] 10 U.S.C. §2157.

[517] NDU is funded under "Defense-wide Operations and Maintenance" as per *NDAA FY 2002*. Service hosts provide infrastructure for NDU at Fort Leslie J. McNair (Army) and Joint Forces Staff College (Navy). The services pay O&M and infrastructure for their own PME schools. Chief, Joint Education and Doctrine Division, JCS/DJ-7, email, 18 March 2010.

[518] Sam Retherford, Briefing, 20 March 2009, and Telephone Interview, 9 November 2009.

[519] *DOD Directive 5124.02*, USD(P&R), June 23, 2008, 4 and 8. See also 10 U.S.C. §661-668.

[520] Office of the Secretary of Defense, *Guidance for the Development of the Force*, and briefings by USD(P) and USD(P&R), 27 August 2009. "The GDF considers a 20-year view of the security environment to inform the construction of the Pentagon's fiscal year 2010-2015 spending plan, according to Jason Sherman," *Inside the Pentagon*, 15 May 2008, 1. This is a classified document but has a short portion related to PME that is unclassified and focuses exclusively on the USD(P&R) role. Also, Sam Retherford, Interview, 9 November 2009. Admiral Michael Mullen, HASC O&I Member Meeting, 27 October 2009. The Subcommittee requested the GDF, but the Office of the Secretary of Defense turned down the request per letter from William Lynn, Deputy Secretary of Defense, 19 October 2009.

[521] Sam Retherford, Briefing, 20 March 2009, and Telephone Interview, 9 November 2009. The Office of the Secretary of Defense Historian's Office provided information that the Deputy Assistant Secretary of Defense for Education existed in the 1960s. In addition to other responsibilities, this office was responsible for PME. A Ph.D. Principal Assistant for Education Programs and Management Training was subordinate to the DASD.

[522] *Ibid.*

[523] *Ibid.*

[524] RADM Admiral Garry E. Hall, MajGen Robert P. Steel, RADM James P. Wisecup, and MG Robert M. Williams,
Responses to Questions for the Record, Congress, U.S. House of Representatives, Committee on Armed Services, Hearing Before the Subcommittee on Oversight and Investigations, *Thinkers and Practitioners: Do Senior Professional Military Education Schools Produce Strategists?* 4 June 2009.

[525] 10 U.S.C. § 153(a)(5)(C).

[526] *OPMEP*, D-4 - D-6.

[527] *Ibid.*, D-5 - D-7.

[528] JCS/DJ-7 Staff, Briefing to the HASC Staff, 17 March 2009. See also Dr. Richard Kohn, Testimony and Responses to Questions for the Record, HASC O&I Subcommittee Hearing, *Another Crossroads?*

[529] Dr. Alexander Cochran, Testimony and Response to Questions for the Record, HASC O&I Subcommittee Hearing, *Another Crossroads?*

[530] JCS/DJ-7, HASC Staff briefing, 17 March 2009.

[531] *OPMEP*, Enclosure C.

[532] USAF Briefing to HASC Staff, 26 March 2009, and HASC Staff Visit to Air University, 8-9 April 2009.

[533] Cochran, HASC Written Testimony, O&I Subcommittee Hearing, *Another Crossroads*, 3.

[534] Breslin-Smith, Written Testimony, HASC O&I Subcommittee Hearing, *Another Crossroads?* 3.

[535] MG Robert Scales (Ret.) and Professor Williamson Murray, as well as serving and retired senior officers in nonattribution interviews.

[536] *Ibid.*

[537] *Ibid.* and HASC Chairman Ike Skelton has also posed this question.

[538] *Report of the Panel on Military Education of the 100th Congress of the Committee on Armed Services*, U.S. House of Representatives, 21 April 1989, (*The Skelton Report*), 108.

[539] *Ibid.*

[540] Cochran, Testimony, HASC O&I Subcommittee Hearing, *Another Crossroads?* See also, HASC O&I Subcommittee Hearing, *Thinkers and Practitioners*. See for example MG Williams, Testimony, HASC O&I Subcommittee Hearing, *Thinkers and Practitioners*. Also, interviews with leaders of various JPME and PME institutions, March – October 2009.

[541] *Ibid.*

[542] David A. Tretler, "Professional Military Education: Background and Basics," HASC O&I Subcommittee Members Briefing, 24 March 2009. Also CODELs Snyder to ICAF and National, 31 March and 29 April 2009, respectively, and HASC Staff Visit to National, 3 April 2009. This sentiment was repeated by other NDU staff and ICAF and National faculty during interviews and staff visits.

[543] CODELs Snyder to ICAF and National, 31 March and 29 April 2009, respectively, and HASC Staff Visit to National, 3 April 2009. Breslin-Smith, Testimony and Response to Questions for the Record, HASC O&I Subcommittee Hearing, *Another Crossroads?* See also HASC O&I Staff Interviews with Faculty and Staff, February – December 2009.

[544] Breslin-Smith, Testimony, HASC O&I Subcommittee Hearing, *Another Crossroads?* Also, Faculty Panels, CODEL Snyder to National, 29 April 2009, and HASC Staff Visit to National, 3 April 2009. See also HASC O&I Staff Interviews with Faculty and Staff, February – December 2009.

[545] CODELs Snyder to ICAF and National, 31 March and 29 April 2009, respectively, and HASC Staff Visit to National, 3 April 2009. HASC O&I Staff Interviews with Faculty and Staff, February – December 2009.

[546] CODELs Snyder to ICAF and National, 31 March and 29 April 2009, respectively and HASC Staff Visits to National, 3 April 2009; MCU, 27 March 2009; and, USAWC, NWC, and AU, 5-9 April 2009. Also, JCS/DJ-7, USD (P&R), and Defense Language Office, Marine Corps, Air Force, Army, and Navy Staffs, HASC Staff Briefings, 17, 20, 23, 26, 26, and 31 March 2009, respectively.

[547] *Ibid.*

[548] Frank Pagano, Capstone Briefing to HASC Staff, 19 August 2009, and RDML Steve Pietropaoli, O&I Staff Interview, 9 October 2009.

[549] NDU-P Interviews, 7 August and 29 October 2009, and Pietropaoli Interview, 9 October 2010. Also, Frank Pagano, Capstone Briefing to HASC Staff, 19 August 2009, and LTG David Barno Interview, 9 March 2010. See also, Paula Thornhill, "Improving Capstone: Change Course Focus to Challenge Participants," *Armed Forces Journal* (online edition), April 2010. One student survey cited the longest serving mentor as "the best one." Other experts say the length of time a senior mentor serves is not important if they stay current in their joint, interagency, intergovernmental, and multinational knowledge.

[550] National Faculty Panel, CODEL Snyder, 29 April 2009; Faculty Panel, Staff Visit to National, 3 April 2009; and Breslin-Smith, Cochran, Dr. James Carafano, and Kohn, Testimony, HASC O&I Subcommittee Hearing, *Another Crossroads.*

[551] HASC CODEL to MCU, 20 May 2009; HASC Staff Visit to MCU, 27 March 2009; HASC Staff Visits to USAWC, NWC, and AU, 5-9 April 2009. Also, Mullen, HASC O&I Member Meeting, 27 October 2009; GEN George Casey, 21 October 2009; Gen Norton Schwartz, 1 October 2009; ADM Gary Roughead, 29 September 2009; and Gen James Conway, 24 September 2009.

[552] U.S. Army Briefings to HASC Staff, 26 March and 17 July 2009; HASC Staff Visits to USAWC 6-7 April 2009, CGSC 27-28 May 2009, USMA 30 June – 1 July 2009, and TRADOC 30 April – 1 May 2009; HASC O&I Subcommittee Hearing, *Raising Thinking from the Tactical to the Operational Level: JPME I and II at the Services' and Joint Command & Staff Colleges*, 24 June 2009; Congress, U.S. House of Representatives, Committee on Armed Services, Hearing Before the Subcommittee on Oversight and Investigations, *Thinkers and Practitioners;* and, Congress, U.S. House of Representatives, Committee on Armed Services, Hearing Before the Subcommittee on Oversight and Investigations, *Beyond Service Core Competency: Are Junior Officers Prepared for Today's Security Environment?* 15 July 2009; HASC O&I Member Meetings with MG Williams, 3 June 2009; LTG Caldwell, 17 June 2009; and, GEN Casey, 21 October 2009; and, HASC O&I Staff Meetings, BG Edward Cardon, 17 July and 17 August 2009, and General Martin Dempsey, 17 September 2009.

[553] Air Force Briefing to HASC Staff, 26 March 2009; HASC Staff Visits to AU, 8-9 April 2009 and to USAFA, 28-29 June 2009; HASC O&I Subcommittee Hearing, *Raising Thinking*; HASC O&I Subcommittee Hearing, *Thinkers and Practitioners;* and, HASC O&I Subcommittee Hearing, *Beyond Service Core Competency*; and, HASC O&I Subcommittee Member Meetings with LtGen Peck, 12 May 2009; LtGen Gould, 15 September 2009; and, Gen Schwarz, 1 October 2009.

[554] Navy Briefings to HASC Staff 31 March and 11 May 2009; CODEL Snyder to USNA, 9 June 2009; HASC Staff Visit to NWC 7-8 April 2009; HASC O&I Hearings, *Thinkers and Practitioners* and *Beyond Service Core Competency*; and, HASC O&I Member Meetings with ADM Roughead, 29 September 2009, and RADM Wisecup, 28 April 2009. Also HASC O&I Subcommittee Hearing, *Thinkers and Practitioners.*

[555] USMC Briefing to HASC Staff 23 March 2009, CODEL Snyder to MCU, 20 May 2009; HASC Staff Visit to MCU, 27 March; HASC O&I Subcommittee Hearings, *Thinkers and Practitioners*; *Raising Thinking*, and *Beyond Service Core Competency*; and, HASC O&I Member Meetings with Gen Conway, 24 September 2009, and MajGen Robert Neller, 5 November 2009.

[556] Carafano, Testimony, HASC O&I Subcommittee Hearing, *Another Crossroads?*

[557] Kohn, Testimony, *Ibid.*

[558] Carafano, Testimony, *Ibid.*

[559] U.S. Army Combined Arms Center, *A Leader Development Strategy for an Expeditionary Army*, Draft 16 June 2009 (revised).

[560] Breslin-Smith, Testimony, HASC O&I Subcommittee Hearing, *Another Crossroads?* See also, McCausland and Weissman, 298-299. Former and current USAWC Board of Visitor Members and USAWC Senior Leaders, HASC Staff Interviews, May 2009.
[561] McCausland and Weissman, 298-299.
[562] *Ibid.*, 299.
[563] HASC Staff Visit to AU, 8-9 April 2009.
[564] Kohn, Written Testimony, HASC O&I Subcommittee Hearing, *Another Crossroads?* 12-13.
[565] HASC Staff Visit to AU, 8-9 April 2009.
[566] Chairman of the Joint Chiefs of Staff, *Process for Accreditation of Joint Education (PAJE): Report of Accreditation of the Air War College*, 20-23 October 2008, 3.
[567] Air War College Dean, Interview with HASC Staff, 9 April 2009. See also, MajGen Maurice Forsyth, Responses to Questions for the Record, *Thinkers and Practitioners*, 4 June 2009, CHARRTS No.: HASCOI-05-006.
[568] McCausland and Weissman, 347.
[569] AWC Dean, Interview with HASC Staff, 9 April 2009, and Faculty Panels, HASC Staff Visit to AU, 9 April 2009.
[570] *The Skelton Report*, 133.
[571] Ibid.
[572] Dr. John Yaeger, email, 10 April 2010. Karen DiFulgo, Vice President, Human Resources, Gaylord National Resort & Convention Center, Interview, 12 April 2010. Dr. Helen Meisenhelder, Organizational Behavior Specialist and Former HQ USAF Personnel Specialist, email to HASC O&I staff, 13 April 2010. See also various sources including, Caroline Sotello Viernes Turner, *Diversifying the Faculty: A Guidebook for Search Committee*, Association of American Colleges and Universities, Washington, D.C. 2002 (with an annotated bibliography) and "Diversifying the Faculty: Search Committee Going Beyond Business as Usual," U.S. Department of Education, Office of Educational Research and Improvement (OERI), National Library of Education. Also, Pamela Bernard, "When Seeking a Diverse Faculty, Watch Out for Legal Minefields," *Chronicle of Higher Education*, Vol. 53, Issue 6, 29 September 2006, 66; Daryl Smith, "How to Diversify the Faculty, *Academe*, American Association of University Professors, 2 March 2007; Daryl Smith et al, "Interrupting the Usual: Successful Strategies for Hiring Diverse Faculty, *The Journal of Higher Education*, Vol. 75, No. 2, April 2004, 133-160; and, Charmaine P. Clowney, J.D., "Best Practices in Recruiting and Retaining Diverse Faculty (slides), PA State System of Higher Education. Also, USMA, *Institutional Self Study Report*, June 2009, 21-45.
[573] Yaeger, email, 10 April 2010 and DiFulgo, Interview, 12 April 2010. Also, Leader and Faculty Panels, CODELs Snyder to ICAF, 31 March 2009, and National, 29 April 2009. Leader and Faculty Panels, HASC Staff Visits to USAFA and USMA, 29 June – 1 July 2009. Leader, Faculty, and Student Panels, HASC Staff Visits to National, 3 April 2009; USAWC, NWC, and AU, 5-9 April 2009; and, CGSC, 26-27 May 2009. See also, Kohn, Testimony, HASC O&I Subcommittee Hearing, *Another Crossroads?* Also, CODEL Snyder to USNA, 9 June 2009. See also, NWC, "Statement of Academic Policy 09-01, Faculty Diversity," 22 May 2009.
[574] Leader and Faculty Panels, CODELs Snyder to ICAF, 31 March 2009, and National, 29 April 2009. Leader and Faculty Panels, HASC Staff Visits to USAFA and USMA, 29 June – 1 July 2009. Leader, Faculty, and Student Panels, HASC Staff Visits to National, 3 April 2009; USAWC, NWC, and AU, 5-9 April 2009; and, CGSC, 26-27 May 2009. Also Kohn, Testimony, HASC O&I Hearing, *Another Crossroads?* Also, illustrative is The Higher Learning Commission (A Commission of the North Central Association of Colleges and Schools), *Report of a Comprehensive Evaluation Visit to the United States Air Force Academy*, "Assurance Section," (USAF Academy, CO: 27-29 April 2009), 11. While 21.5% of cadets are women and 21.3% are racial or ethnic minorities, among the faculty only 16.4% are women and 8.8% are racial or ethnic minorities.
[575] CODEL Snyder to USNA, 9 June 2009; Leader and Faculty Discussions and Panels, Staff Visit to USAFA and USMA, 29 June – 1 July 2009.
[576] NWC, "Statement of Academic Policy 09-01, Faculty Diversity," 22 May 2009.
[577] Faculty Panel, CODEL Snyder, ICAF, 31 March 2009; Faculty Panel, HASC Staff Visit, National, 3 April 2009; and, Faculty Panel, HASC Staff Visit, USAFA, 29 June 2010.
[578] Interviews with BG Edward Cardon, 17 July and 17 August; Harvey Rishikof, 25 July 2009; and Current and Former Faculty Members from Various PME Institutions, March – November 2009. HASC O&I Subcommittee Hearings and Responses to Questions for the Record 4 June – 15 July 2009. HASC Staffdels to National, 3 April 2009; USAWC, NWC, and AU, 5-9 April 2009; CGSC, 27 May 2009; and, USAFA and USMA 29 June – 1 July 2009. The copyright issue was raised at USAWC, NWC, CGSC, USAFA, ACSC, and USMA. Academic freedom was raised specifically at AU, USAFA, and USMA.
[579] *Skelton Report*, 133.
[580] *Ibid.*, 147.
[581] *Ibid.*, 148.
[582] *Ibid.*
[583] HASC O&I Subcommittee Hearing, *Another Crossroads?*
[584] Kohn, Response to Questions for the Record, HASC O&I Subcommittee Hearing, *Another Crossroads?*

[585] HASC Staff Interviews with Various Current and Former Faculty Members on CODELs and Staff Visits, and individually, February-December 2009.

[586] Murray, Written Testimony, Congress, U.S. House of Representatives, Committee on Armed Services, Hearing Before the Subcommittee on Oversight and Investigations, *Charting the Course for Effective Professional Military Educaiton*,10 September 2009, 23.

[587] *RADM* Wisecup, BG Cardon, and ACSC, Responses to Questions for the Record, HASC O&I Subcommittee Hearing, *Raising Thinking*, CHARRTS No.: HASC OI-06-12.

[588] BG Kasun, RADM Wisecup, Col Damm, BG Cardon, and ACSC, Responses to Questions for the Record, HASC O&I Subcommittee Hearing, *Raising Thinking*, CHARRTS No.: HASC OI-06-12. See also: MajGen Steel, RADL Hall, and MajGen Forsyth, Testimony, HASC O&I Subcommittee Hearing, *Thinkers and Practitioners*.

[589] HASC O&I Subcommittee Hearing, *Raising Thinking* and HASC O&I Subcommittee Hearing, *Thinkers and Practitioners*.

[590] *Ibid.*

[591] LtGen Paxton, Written Testimony, Congress, U.S. House of Representatives, Committee on Armed Services, Hearing Before the Subcommittee on Oversight and Investigations, *Investing in Our Military Leaders: The Role of Professional Military Education in Officer Development*, 8.

[592] *OPMEP*.

[593] *The Skelton Report*, 148-149.

[594] McCausland and Weissman, 6.

[595] Cochran, Testimony and Response to Questions for the Record, HASC O&I Subcommittee Hearing, *Another Crossroads?*

[596] Breslin-Smith, Written Testimony, HASC O&I Subcommittee Hearing, *Another Crossroads?* 4-5.

[597] Cochran, Response to Questions for the Record, HASC O&I Subcommittee Hearing, *Another Crossroads?*

[598] Kohn, Written Testimony, HASC O&I Subcommittee Hearing, *Another Crossroads?* 12.

[599] RADL Hall, Written Testimony, HASC O&I Subcommittee Hearing, *Thinkers and Practitioners*, 19.

[600] BG Kasun and Col Damm, Responses to Questions for the Record, HASC O&I Hearing, *Raising Thinking*, CHARRTS No.: HASC OI-06-012.

[601] *The Skelton Report*, 133. See also, Andrew Krepinevich, Memo to Andrew Marshall, "Rebuilding America's Intellectual Arsenal, 18 December 2005. Krepinevich asked, among other questions, "What is being done to attract top faculty members to our key military education centers?"

[602] *The Skelton Report*, 133.

[603] Senior Leader and Faculty Panels, CODELs Snyder to ICAF, 31 March 2009; National, 29 April 2009; and MCU, 20 May 2009; and, National Senior Leader Panel for HASC O&I Members, 9 June 2009. Also, Leader and Faculty Panels, HASC Staff Visits to National, 3 April 2009; USAWC, NWC, and AU, 5-9 April 2009; JFSC, 30 April – 1 May 2009; and, CGSC, 26-27 May 2009. Discussions concerned the need to gap military positions taking who they are offered so they will not have to gap positions. Discussions also covered the issue of not enough operators (combat arms, pilots, or line officers) being available to attend or serve as faculty for PME courses.

[604] Cochran and Kohn, Written Testimony, HASC O&I Subcommittee Hearing, *Another Crossroads?* 3 and 10, respectively. See also school senior leader and faculty interviews, February-December 2009.

[605] Representative Steve Israel, Notes provided from the "Cognitive Transformation Roundtable," United States Capitol, Washington D.C., 16 February 2005. Attendees included Rep. Ike Skelton, MG Robert Scales, and a number of other senior military leaders, academics, and policy makers.

[606] *The Skelton Report*, 136.

[607] *Ibid.*, 136.

[608] *Ibid.*, 146.

[609] 5 U.S.C. §5101-5115 and 10 U.S.C. §§1595, 4021, 7478, and 9021. For the service academies, 5 U.S.C. §5102(c)(10) and 10 U.S.C. Chapters 403, 603, and 903.

[610] *The Skelton Report*, 32, 41, and 146.

[611] HASC Staff Visit to NWC, 6-9 April 2009. As an example, NWC Provost requested a waiver from the State Department for retired Ambassador Peters, and was turned down in 2008. See 5 U.S.C. §8344 and Teddy B. Taylor, Acting Director General of the Foreign Service and Director of Human Resources, Letter to RADL Jacob L. Shuford, President of the NWC, 20 August 2008. Other PME faculties and senior staffs raised this issue with HASC staff and Subcommittee Members in various fora in discussing increasing the number of interagency faculty.

[612] *The Skelton Report*, 135, 142, 146, and 147. Also, HASC Staff visits to NWC, 3 April 2009; USMC University, 27 March 2009; and, USAWC, NWC, and AU, 5-9 April 2009. Also JCS/DJ-7, Marine Corps, Air Force, Army, Navy and Navy (N15), Briefings to the HASC Staff, 17 March, 23 March, 26 March, 26 March, 31 March, and 11 May 2009, respectively.

[613] BG Cardon, BG Kasun, RADM Wisecup, Col Damm, and ACSC Staff, Responses to Question for the Record, HASC O&I Subcommittee Hearing, *Raising Thinking*, CHARRTS No.: HASC OI-06-002.

[614] HASC O&I Subcommittee Hearing, *Investing in Our Military Leaders*. Also, Briefings to HASC staff by JCS/DJ-7, 17 March; USAF, 26 March; Army, 26 March; Navy, 31 March; and, Navy N-15, 11 May 2009.

[615] *The Skelton Report*, 32, 41, 110, and 128.

[616] BG Kasun, Response to Question for the Record, HASC O&I Subcommittee Hearing, *Raising Thinking*, CHARRTS No.: HASC OI-06-002.

[617] BrigGen Dana Born and BG Patrick Finnegan, Responses to Question for the Record, Congress, U.S. House of Representatives, Committee on Armed Services, Hearing Before the Subcommittee on Oversight and Investigations, *Beyond Service Core Competency,: Are Junior Officers Prepared for Today's Security Environment?* CHARRTS No.: HASCOI-08-034 and HASCOI-08-009, respectively.

[618] HASC Staff Visit to MCU, 27 March 2009.

[619] CODEL Snyder and HASC Staff Visit to MCU, 20 May and 27 March 2009, respectively.

[620] Caldwell, Responses to Question for the Record, HASC O&I Subcommittee Hearing, *Investing in Our Military Leaders*, CHARRTS No.: HASCOI-07-011.

[621] Sitterly, Responses to Question for the Record, HASC O&I Subcommittee Hearing, *Investing in Our Military Leaders*, CHARRTS No.: HASCOI-07-039.

[622] Lutterloh, Responses to Question for the Record, HASC O&I Subcommittee Hearing, *Investing in Our Military Leaders*, CHARRTS No.: HASCOI-07-021.

[623] AWC Staff, Response to HASC Questions, 22 and 23 June 2009. BG Cardon, BG Kasun, RADM Wisecup, Col Damm, and ACSC Staff, Responses to Question for the Record, HASC O&I Hearing, *Raising Thinking*, CHARRTS No.: HASC OI-06-002.

[624] 10 U.S.C. §619.

[625] *National Defense Authorization Act for Fiscal Year 2007*, P.L. 109-364, Section 519, 17 October 2006, amending 10 U.S.C. §668.

[626] *Memorandum of Agreement: Assignment of Military Faculty at Senior Service Level Colleges*, 30 December 2005.

[627] LtGen Paxton, LTG Caldwell, Lutterloh, and MajGen Spiese, Testimonies and Written Testimonies, HASC O&I Subcommittee Hearing, *Investing in Our Military Leaders*, 28 July 2009, 9, 9, 10, and 16-17, respectively.

[628] *National Defense Authorization Act for Fiscal Year 2007*, PL 109-364, 17 October 2006, 120 STAT. 2190, §519, 108-109.

[629] CGSC Information Paper coordinated with the Air Force, Navy, and Marine Corps Elements at Fort Leavenworth, "House Armed Services Committee (HASC) Question," 12 May 2009, 2-3. Also, BG Cardon, Written Testimony and Response to Question for the Record, HASC O&I Hearing, *Raising Thinking*, 11, and CHARRTS No.: HASCOI-06-007. See also BG Caldwell, Testimony, HASC O&I Subcommittee Hearing, *Investing in Our Military Leaders*.

[630] Reference to the ILE Faculty MOU by BrigGen Melvin Spiese, Responses to Question for the Record, HASC O&I Subcommittee Hearing, *Investing in Our Military Leaders*, CHARRTS No.: HASCOI-07-030. BG Cardon, Response to Question for the Record, HASC O&I Subcommittee Hearing, *Raising Thinking*, CHARRTS No.: HASCOI-06-007. See also LTG Caldwell, Testimony, HASC O&I Subcommittee Hearing, *Investing in Our Military Leaders*.

[631] LtGen Paxton, Written Testimony, HASC O&I Subcommittee Hearing, *Investing in Our Military Leaders*, 9. See also LTG Caldwell, Written Testimony, 9; Lutterloh, Written Testimony, 10; and MajGen Spiese, Written Testimony, 16. PAJE Accreditation Reports for Air Command and Staff College (2007), Enclosure A, 3, and Command and General Staff College (2008), 2 and 6. RADM Wisecup, Col Damm, BG Cardon, and ACSC Staff, Responses to Question for the Record, HASC O&I Subcommittee Hearing *Raising Thinking*, CHARRTS No.: HASCOI-06-007 and BG Cardon, HASCOI-06-003. MCU, "Joint Duty Assignment List (JDAL) Billets at JPME I Institutions, Submitted in Response to Questions Raised by HASC O&I, 28 July 2009," 17 August 2009. U.S. Army Combined Arms Center letter to the House Armed Services Committee, 20 August 2009. Faculty and Senior Leader Panels and Interviews, Staff visits to MCU, 27 March 2009; AU, 8-9 April; and CGSC, 27 May 2009. Cardon, HASC Staff Interview, 17 July 2009. MajGen Neller, HASC O&I Member Meeting, 5 November 2009. Dr. Jerre Wilson, Emails, MCU, November 2009.

[632] *Fiscal Year 2010 National Defense Authorization Bill, House Report 111-166*, 18 June 2009, 315. See also, Snyder-Wittman letter to ADM Michael Mullen, 23 July 2009. USD(P&R) and JCS/DJ-1 (Personnel) were to have a draft report in coordination on or about 1 April 2010, per emails from JCS/DJ-7, 18 March 2010, and OSD Legislative Liaison, 26 March 2010.

[633] *Fiscal Year 2010 National Defense Authorization Bill, Conference Report 111-288*, 7 October 2009, 726. See also, Joint Staff Letter to the Chairman, Oversight and Investigations Subcommittee of the House Armed Services Committee, "Response to Joint Professional Military (JPME) I Faculty on Joint Duty Assignment List (JDAL)," 17 September 2009.

[634] HASC O&I Subcommittee Chairman Snyder and Col Stephen Tanous exchange during HASC O&I Subcommittee Hearing, *Beyond Service Core Competency*.

[635] 10 U.S.C. §624.

[636] Department Head Panel, HASC Staff Visit, USMA, 1 July 2009.

[637] ADM Mullen, (ADM Pat Walsh (Blue Angels pilot, Ph.D., and USNA faculty member) among others), HASC O&I Meeting, 27 October 2009. In addition, according to the USAWC Commandant, part of the complicating factor in terms of time is that Army command positions that in the past lasted two years, are now lasting three or longer. Having longer command tours, and even early promotions due to attrition, reduces time available for PME, faculty duty, joint positions, advanced education, and other possible career broadening assignments.

[638] Under Secretary of Defense for Personnel and Readiness (USD(P&R)), *Department of Defense Instruction 1402.06, Civilian Faculty Positions in Department of Defense (DoD) Post-Secondary Educational Institutions*, 6 November 2007.

[639] LtGen Frances Wilson, NDU President, Telephone Interview, 28 April 2009.

[640] BG Kasun, Response to Questions for the Record, *Raising Thinking*, CHARRTS No.: HASCOI-06-013.

[641] ACSC Staff, Response to Questions for the Record, HASC O&I Subcommittee Hearing, *Raising Thinking*, CHARRTS No.: HASCOI-06-013.

[642] Col Damm, Response to Questions for the Record, HASC O&I Subcommittee Hearing, *Raising Thinking*, CHARRTS No.: HASCOI-06-013.

[643] BG Cardon, Response to Questions for the Record, HASC O&I Subcommittee Hearing, *Raising Thinking*, CHARRTS No.: HASCOI-06-013. BrigGen Born, Response to Question for the Record, HASC O&I Subcommittee Hearing, *Beyond Service Core Competency*, CHARRTS No.: HASCOI-08-038. Faculty Panel, Staff Visit to USAFA, 29 June 2009.

[644] RADM Wisecup, Response to Questions for the Record, HASC O&I Subcommittee Hearing, *Raising Thinking*, CHARRTS No.: HASCOI-06-013.

[645] Col Belcher, MajGen Forsyth, MG Williams, RADM Wisecup, MajGen Steel, and RADM Hall, HASC O&I Hearing, Responses Question for the Record, *Thinkers and Practitioners*, CHARRTS No.: HASCOI-05-005, 4 June 2009.

[646] HASC Staff Visits to AWC, NWC, and AU 6-9 April 2009; and to CGSC 27 May 2009. RADM Wisecup and BG Cardon, Written Testimony. HASC O&I Hearing, *Raising Thinking*.

[647] Murray, Written Testimony, HASC O&I Subcommittee Hearing, *Charting the Course*, 10 September 2009, 23.

[648] BrigGen Born, Response to Question for the Record HASC O&I Subcommittee Hearing, *Beyond Service Core Competency,*, CHARRTS No.: HASCOI-08-038.

[649] Capt Klunder, Response to Question for the Record, HASC O&I Subcommittee Hearing, *Beyond Service Core Competency*, CHARRTS No.: HASCOI-08-058.

[650] ACSC Staff, Response to Questions for the Record, HASC O&I Subcommittee Hearing, *Raising Thinking*, CHARRTS No.: HASCOI-06-013.

[651] BG Cardon, Response to Questions for the Record, HASC O&I Subcommittee Hearing, *Raising Thinking*, CHARRTS No.: HASCOI-06-013.

[652] Col Damm, Response to Questions for the Record, HASC O&I Subcommittee Hearing, *Raising Thinking*, CHARRTS No.: HASCOI-06-013.

[653] BG Kasun, Response to Questions for the Record, HASC O&I Subcommittee Hearing, *Raising Thinking*, CHARRTS No.: HASCOI-06-013.

[654] RADM Wisecup, Response to Questions for the Record, HASC O&I Subcommittee Hearing, *Raising Thinking*, CHARRTS No.: HASCOI-06-013, and *Thinkers and Practitioners?* CHARRTS No.: HASCOI-05-017, on "indefinite appointments."

[655] Capt Klunder, Response to Question for the Record, HASC O&I Subcommittee Hearing, *Beyond Service Core Competency*, CHARRTS No.: HASCOI-08-054.

[656] Senior Leader and Faculty Panels, CODEL Snyder to USNA, and HASC Staff Visit to USNA, 9 June 2009.

[657] Cardon, Response to Question for the Record, HASC O&I Subcommittee, Hearing, *Raising Thinking*, CHARRTS No.: HASCOI-06-009.

[658] HASC Staff Interviews with National Faculty, March 2009. Also, according to Judith Hicks Stiehm, the turn over came in 1991 and faculty members were given a choice whether to maintain Title 5 status or to convert to Title 10 appointments, *The U.S. Army War College: Military Education in a Democracy*, (Temple University Press: Philadelphia, PA, 2002), 76.

[659] LtGen Frances Wilson, NDU-P, Telephone Interview, 9 June 2009.

[660] 10 U.S.C. §§2156, 4021, 7478, and 9021.

[661] Senior Leader and Faculty Panels, HASC Staff Visits to MCU, 27 March 2009, and Naval War College, 7-8 April 2009.

[662] Faculty handbooks including CGSC, *Title 10 Civilian Faculty Manual CGSC Pamphlet 690-1*, August 2008, and MCU, *Title 10 Faculty Handbook*, June 2008. *National Defense University (NDU) Regulation 690-4, Personnel-Civilian, Employment under 10 USC §1595*, 4 August 2005.

[663] Faculty Panels, CODELs Snyder to ICAF, 29 April 2010, and National, 3 April 2009; Staff Visits, USAWC, NWC, and AU, 5-9 April 2009; Staff Visit, CGSC, 27-29 May 2009; and, Staff Visits, USAFA and USMA, 29 June – 1 July 2009.

[664] Appointment letters or memoranda including undated example from NWC, ICAF examples from 21 June 2006 and 21 May 2008, ACSC example dated 21 July 2009, and an undated example form letter and job

announcement from AWC. Faculty handbooks including CGSC, *Title 10 Civilian Faculty Manual CGSC Pamphlet 690-1*, August 2008 and MCU, *Title 10 Faculty Handbook*, June 2008. Some faculty members from National reported that either they did not receive a formal appointment letter or could not locate it.

[665] *National Defense University (NDU) Regulation 690-4, Personnel-Civilian, Employment under 10 U.S.C. §1595*, 4 August 2005. HASC Staff Visit to National, 3 April 2009. See also Staff Discussion with ICAF Faculty, 2 June 2009.

[666] MajGen Steel, Response to Question for the Record, HASC O&I Subcommittee Hearing, *Thinkers and Practitioners*, CHARRTS No.: HASCOI-05-014.

[667] AWC Staff, HASC Staff Interview, 9 April 2009. See also Forsyth, Response to Question for the Record, HASC O&I Subcommittee Hearing, *Thinkers and Practitioners*, 6.

[668] BG Cardon, Response to Question for the Record, HASC O&I Subcommittee Hearing, *Raising Thinking*, CHARRTS No.: HASCOI-06-009.

[669] Col Damm and Col Belcher, Response to Question for the Record, HASC O&I Subcommittee Hearing, *Raising Thinking*, CHARRTS No.: HASCOI-06-009, and *Thinkers and Practitioners*, CHARRTS No.: HASCOI-05-006, respectively. See also MCU, *Title 10 Faculty Handbook*, June 2008. Although both Col Damm and Col Belcher address this issue, the quote is specifically from Col Belcher's 4 June response. RADM Wisecup, Response to Question for the Record, HASC O&I Subcommittee Hearing, *Raising Thinking*, CHARRTS No.: HASCOI-06-009 and -016. Also MG Williams, Response to Question for the Record, HASC O&I Subcommittee Hearing, *Thinkers and Practitioners*, CHARRTS No.: HASCOI-05-006. See also *Carlisle Barracks Memorandum 690-2* "Civilian Personnel: Employment Under Title 10 Code Section 4021," 29 March 2009.

[670] MajGen Forsyth, HASC Response to Question for the Record, HASC O&I Subcommittee Hearing, *Thinkers and Practitioners*, CHARRTS No.: HASCOI-05-006. See also *Spaatz Center for Officer Education Operating Instruction 36-3*, "Faculty Management" and "Air War College Supplement" to AFI 36-804.

[671] BG Kasun, Response to Question for the Record, HASC O&I Subcommittee Hearing, *Raising Thinking*, CHARRTS No.: HASCOI-06-009. See also *NDU-P Regulation 690-4*, paragraphs 8b and 12b (2).

[672] MajGen Steel and RADM Hall, Response to Question for the Record, HASC O&I Subcommittee Hearing, *Thinkers and Practitioners*, CHARRTS No.: HASCOI-05-006. *NDU-P Regulation 690-4*, Appendix C paragraph 7, 4 August 2005. Faculty Interviews, March – November 2009. Senior Leader and Faculty Panels, CODEL Snyder to ICAF, 31 March 2009. Faculty Panels, CODEL Snyder, and HASC Staff Visit to National, 29 and 3 April 2009, respectively. National Senior Leader Panel, HASC O&I Members, 9 June 2009. ICAF Faculty Handbook, 2008.

[673] Faculty Panel, HASC Visit to USMA, 1 July 2009.

[674] Senior Leader and Faculty Panels, CODEL Snyder to ICAF, 31 March 2009, and ICAF Faculty and Senior Leader Interviews with HASC Staff, 2 June, 29 July, and 12 August, 2009.

[675] CJCS, *Process for Accreditation of Joint Education Reaffirmation of the Industrial College of the Armed Forces (ICAF)*, 6-11 February 2005, 7-9.

[676] Senior Leader and Faculty Panels, CODELs Snyder, ICAF 31 March 2009, National 29 April 2009, MCU 20 May 2009, USNA 9 June 2009. LtGen Wilson, LtGen Peck, MajGen Steel, and LTG Caldwell Member Meetings, 24 March 2009, 25 March 2009, 12 May 2009, 3 June 2009, and 18 June 2009, respectively. HASC Staff Visits and Interviews including NDU-P and NDU Senior Staff, 16 March 2009; Senior Leader and Faculty Panels, MCU 27 March 2009; Senior Leader and Faculty Panels, HASC Staff Visits, USAWC, NWC, and AU, 6-9 April 2009; National Senior Leader and Faculty Panels 3 April 2009; RADM Wisecup Interview, 28 April 2009; JFSC Senior Leader and Faculty Panels, HASC Staff Visits, 30 April – 1 May 2009; ICAF Faculty Interview, 2 June 2009; LTG Caldwell Interview, 17 June 2009; Senior Leader and Faculty Panels, USAFA and USMA, 29 June – 1 July 2009; BG Cardon Interview, 17 July 2009; and, ICAF Dean and Associate Dean Interview, 29 July 2009.

[677] NWC Faculty Panel, HASC Staff Visit, 6-7 April 2009.

[678] USAFA Senior Leader and Faculty Panels, HASC Staff Visit, 28-29 June 2009. See also BrigGen Born, Response to Question for the Record, HASC O&I Subcommittee Hearing, *Beyond Service Core Competency*, CHARRTS No.: HASCOI- 08-035. See also *A Commission of the North Central Association of Colleges and Schools*, "Statement of Affiliation Status: United States Air Force Academy," 28 October 2009. USAFA Faculty Interviews and Correspondence, HASC staff, June – December 2009. See also Admiral Charles R. Larson for the Secretary of the Air Force, *Study and Report Related to Permanent Professors at the United Stated Air Force Academy*, April 2004, 2-14 and D-2.

[679] USMA Senior Leaders Panel, HASC Staff Visit, 1 July 2009.

[680] Larson, 2-14 and D-2. See also USNA, *Faculty Handbook*, 1998.

[681] *Ibid.*, 3-4 and 3-5.

[682] Faculty Panels, CODELs Snyder, ICAF 31 March 2009 and National 29 April 2009. Also Senior Leader and Faculty Panel HASC Staff Visit, National, 3 April 2009. Tretler, HASC O&I Member Meeting, 24 March 2009. HASC Staff, Faculty Interviews, HASC Staff, March – December 2009.

[683] Faculty Panels, CODELs Snyder, ICAF 31 March 2009 and National 29 April 2009. Senior Leader and Faculty Panels, HASC Staff Visits, National, 3 April 2009; USAWC, 6-7 April 2009; and, USMA and USAFA 29 June – 1 July 2009. Tretler, HASC O&I Member Meeting, 24 March 2009. Faculty Interviews, HASC Staff, March – December 2009.

[684] Faculty Panel, CODEL Snyder, National, 29 April 2009. Senior Leader and Faculty Panels, HASC Staff Visits, National, 3 April 2009; USAWC, 6-7 April 2009; and USMA and USAFA 29 June – 1 July 2009. Tretler, HASC O&I Member Meeting, 24 March 2009. Faculty Interviews, HASC Staff, March – December 2009.

[685] Faculty Panels, CODELs Snyder, ICAF, 31 March 2009 and National, 29 April 2009. Senior Leader and Faculty Panels, HASC Staff Visits, National, 3 April 2009 and USMA and USAFA 29 June – 1 July 2009. Tretler, HASC O&I Member Meeting, 24 March 2009. Faculty Interviews, HASC Staff, March – December 2009.

[686] Faculty Panels, CODELs Snyder, ICAF 31 March 2009 and National 29 April 2009. Senior Leader and Faculty Panels, HASC Staff Visit, National, 3 April 2009. Tretler, HASC O&I Member Meeting, 24 March 2009. Faculty Interviews, March – December 2009.

[687] Faculty Panels, CODELs Snyder, ICAF 31 March 2009 and National 29 April 2009. Senior Leader and Faculty Panels, HASC Staff Visits, National, 3 April 2009; AU, 8-9 April 2009; and USMA and USAFA 29 June – 1 July 2009. Tretler, HASC O&I Member Meeting, 24 March 2009. Also Faculty Interviews, HASC Staff, March – December 2009.

[688] Kohn, Testimony, HASC O&I Subcommittee Hearing, *Another Crossroads*. USMA, *Academic Freedom Advisory Committee Report*, 22 May 2009. Faculty Panels, CODELs Snyder, ICAF 31 March 2009 and National 29 April 2009. Senior Leader and Faculty Panels, HASC Staff Visits, National, 3 April 2009; USAWC, NWC, and AU, 6-9 April 2009; JFSC, 30 April – 1 May 2009; and USMA and USAFA 29 June – 1 July 2009. HASC O&I Member Meeting, 24 March 2009. Faculty Interviews, HASC Staff, March – December 2009.

[689] USMA Faculty Panel, HASC Staff Visit, 1 July 2009.

[690] Col Belcher, Response to Question for the Record, HASC O&I Subcommittee Hearing, *Thinkers and Practitioners*, CHARRTS No.: HASCOI-05-007.

[691] Dr. Andrew Krepinevich memo to Andrew Marshall, "Rebuilding America's Intellectual Arsenal," 18 December 2005.

[692] RADM Wisecup, Response to Question for the Record, HASC O&I Subcommittee Hearing, *Thinkers and Practitioners*, CHARRTS No.: HASCOI-05-007. MajGen Forsyth and MG Williams, Responses to Question for the Record, HASC O&I Subcommittee Hearing, *Thinkers and Practitioners*, CHARRTS No.: HASCOI-05-007.

[693] ICAF and National Faculty, HASC Staff Interviews, March – November 2009.

[694] Representative Steve Israel, Notes provided from the "Cognitive Transformation Roundtable," United States Capitol, Washington DC, 16 February 2005. Attendees included Rep. Ike Skelton, MG Robert Scales, and a number of senior military leaders, academics, and policy makers.

[695] *Ibid.*

[696] Wisecup and Steel, HASC Staff Interviews, May – June 2009. Senior Leader Panels, HASC Staff visit, USAWC, NWC, and AU, 5-9 April 2009. NWC requested a waiver for Provost, retired Ambassador Peters, under Section 824(g) of the Foreign Service Act, as amended (22 U.S.C. 4064(g)), and was turned down in 2008. Teddy B. Taylor, Acting Director General of the Foreign Service and Director of Human Resources, Letter to RADL Jacob L. Shuford, President of the Naval War College, 20 August 2008. See 5 U.S.C. §8344. See also *DOD Instruction 1400.25*, Volume 300, "DOD Civilian Personnel Management System: Employment of Federal Civilian Annuitants in the Department of Defense," 10 December 2008. Title 5 U.S.C §2105, 3323, 8336, 8337, 8344, 8345, 8414, 8455, 8465, 8468, and 9902. Title 26 USC Subchapter A of chapter 21. Title 5 Code of Federal Regulations Subparts B and G of part 330 and part 553. Government civilians retired under Civilian Service Retirement System (CSRS) and the Federal Employee Retirement System (FERS) can keep their annuities.

[697] *National Defense Authorization Act for FY 2010*, P.L. 111-84, sec. 1121 *House Report 111-166: Fiscal Year 2010 National Defense Authorization Bill*, 18 June 2009, sec. 1107, 397.

[698] 10 U.S.C. §§1595, 4021 [Army War College and Command and General Staff College], 7478 [Navy War College and Marine Corps University], and 9021 [Air Force]. National Defense University is included under Section 1595 but is not subject to the same time limit; presumably this was to enable the then twelve-week AFSC (now 10-week JCWS) to hire under this authority.

[699] MajGen Spiese, Written Testimony, HASC O&I Subcommittee Hearing, *Investing in Our Military Leaders*, 12. See also transcript for exchange between Dr. Snyder, and MajGen Spiese, Lutterloh, and LTG Caldwell. In addition, AU Commander, Paper on U.S.C. 10, §9021, and LtGen Peck, HASC Staff Visit, AU, 9 April 2009.

[700] USMC Information Paper, "Additional Thoughts on PME or JPME Programs," 15 October 2009.

[701] Dr. Jerre Wilson, "Advantages of Title 10 Faculty for Programs under 10 Months in Duration," MCU, 12 November 2009.

[702] *Ibid.*

[703] AU Commander, Paper on "Title 10, Sec 9021," HASC Staff Visit, AU, 9 April 2009.

[704] HASC Military Personnel memo on civilian hiring authorities in 10 U.S.C. §§1746, 7043, 7044, 9314, 4338, 6952, and 903), 5 U.S.C. §5102, and 20 U.S.C., 13 November 2009.

[705] NDAA FY 1994, P.L. 103-160, Sec. 533.

[706] NDAA FY 1993, P.L. 102-484, Sec. 523 adding 10 U.S.C §§4331(c) and 9331 (c), and NDAA FY 1994, P.L. 103-160, which struck the 1993 language in favor of adding 10 U.S.C. §§4338 and 9338.

[707] Faculty Panel, Staff Visit, USAFA, 29 June 2009. P.L. 108-136, NDAA FY 04, Sec. 528 C 2 (c). Larson, 4-1. Dick Cheney (Chair) and Bill Taylor (Project Director and Editor), *Professional Military Education: An Asset for Peace and Progress, A Report of the CSIS Study Group on Professional Military Education*, (The Center for Strategic and International Studies, Washington D.C.: March 1997), 32 and 36.

[708] The Higher Learning Commission, A Commission of the North Central Association of Colleges and Schools, "Advancement Section," *Report of a Comprehensive Evaluation Visit to United States Air Force Academy*, USAF Academy, CO, 27-29 April 2009, 7.

[709] Faculty Panels, HASC Staff Visit, USAFA and USMA, 29 June – 1 July 2009.

[710] Larson, ES-1 and ES-3, 3-4 and 3-5, and 5-2.

[711] Senior Leader and Faculty Panels, HASC Staff Visit to USAFA and USMA, 28 June – 1 July 2009; BrigGen Born, Response to Questions for the Record, HASC O&I Subcommittee Hearing, *Beyond Service Core Competency*, CHARRTS No.: HASCOI-08-037. Faculty and Former Faculty Interviews, HASC Staff, March – November 2009.

[712] *Ibid.*

[713] Larson, 3-4 and 3-5.

[714] Senior Leader and Faculty Panels, HASC Staff Visit to USAFA and USMA, 28 June – 1 July 2009, and BrigGen Born, Response to Questions for the Record, HASC O&I Subcommittee Hearing, *Beyond Service Core Competency*, CHARRTS No.: HASCOI-08-037. Also Faculty and Former Faculty Interviews, HASC Staff, March – November 2009.

[715] P.L. 108-136, NDAA FY 04, sec. 528 C 2 (c).

[716] *Ibid.* See also Larson, 2-18, and Cheney and Taylor, 32 and 36.

[717] Larson, 2-18, ES-1, and discussed on 3-4 and 5-1. See also Cheney and Taylor, 32 and 36.

[718] Larson, ES-1, and discussed on 3-4 and 6-3. See also Cheney and Taylor, 32 and 36.

[719] BrigGen Born, Testimony and Response to Question for the Record, HASC O&I Subcommittee Hearing, *Beyond Service Core Competency*, CHARRTS No.: HASCOI-08-037. Although BrigGen Born responded to QFRs that civilian faculty members have served temporarily as department heads during the permanent professors' deployments or sabbaticals, the Larson Report specifies that USAFA civilian faculty cannot, by USAFA policy, hold Department Head and Deputy Department Head positions (Larson, 2-14). Faculty Panel, HASC Staff Visit, USAFA, 29 June 2009. Faculty and Former USAFA Faculty Members, HASC Staff Interviews, June – August 2009.

[720] BrigGen Born, Testimony and Response to Question for the Record, HASC O&I Subcommittee Hearing, *Beyond Service Core Competency.*

[721] BrigGen Born, Response to Question for the Record, HASC O&I Subcommittee Hearing, *Beyond Service Core Competency*, CHARRTS No.: HASCOI-08-037.

[722] *Ibid.*

[723] Larson, ES-1, see also 3-4 and 6-3. Also Senior Leader and Faculty Panels, HASC Staff Visit, USAFA, 28-29 June 2009. Cheney and Taylor, 32 and 36.

[724] Larson, ES-1, see also 3-4, 3-5, and 6-3. Also Senior Leader and Faculty Panels, HASC Staff Visit, USAFA, 28-29 June 2009.

[725] CGSC Senior Leaders Panel, HASC Staff Visit, 27 May 2009.

[726] CGSC Student Panel, HASC Staff Visit, 27 May 2009.

[727] National Defense University website, Academic Affairs, Academic Policies, "5.00 *Non-Attribution/Academic Freedom," 15 April 2003, http://www.ndu.edu/aa/AcademicFreedom, last accessed 20 November 2009. NDU Regulation 360-1, *Academic Freedom and Public Information*. See also http://www.highered. org/resources/AAUP_1940stat.htm.

[728] *DoD Directive 5230.09*, "Clearance of DoD Information for Public Release," 22 August 2008, paragraph 4.e., 2, (emphasis added).

[729] AAUP, "Freedom in the Classroom," 2007 at http://www.higher-ed.org/resources/AAUP_1940stat.htm.

[730] As quoted by BG Kasun, Response to Question for the Record, HASC O&I Subcommittee Hearing, *Raising Thinking*, CHARRTS No.: HASCOI-06-014. See also, USMA, *Institutional Self Study Report*, June 2009, 32-45, and Middle States Commission on Higher Education, "Final Report to the Faculty, Administration, Trustees, Cadets of the United States Military Academy," 13-16 September 2010.

[731] National Defense University, "Self-Study Design Plan," Submitted to Middle States Commission on Higher Education, 15 March 2010, 12.

[732] Higher Learning Commission, *Report of a Comprehensive Evaluation Visit to United States Air Force Academy*, 27-29 April 2009, 30.

[733] Cochran and Breslin-Smith, Question for the Record Responses, HASC O&I Subcommittee Hearing, *Another Crossroads*.

[734] Jackson A. Niday II and Kathleen Harrington, "Can Academic Freedom Work in Military Academies?" *Academe Online*, American Association of University Professors, 2007, last accessed 28 October 2009.

[735] MG Williams, Response to Questions for the Record, HASC O&I Subcommittee Hearing, *Thinkers and Practitioners*, CHARRTS No.: HASCOI-05-005. Col Belcher, Response to Questions for the Record, HASC O&I Subcommittee Hearing, *Thinkers and Practitioners*, CHARRTS No.: HASCOI-05-005. Also see Damm, Response to Question for the Record, HASC O&I Subcommittee Hearing, *Raising Thinking*, CHARRTS No.: HASCOI-06-014. MCU also allows faculty to respond to press inquiries as long as they include a disclaimer. "We consider requests for our faculty as part of our outreach program, necessary for the academic growth of our faculty as well as a good news story about our University."

[736] ACSC Staff, Response to Question for the Record, HASC O&I Subcommittee Hearing, *Raising Thinking*, CHARRTS No.: HASCOI-06-014.

[737] Faculty Panels, HASC Staff Visit, AU, 9 April 2009. AU Faculty Interviews and Communications, April – November 2009. Faculty Panels, HASC Staff Visit, USAFA and USMA, 28 June – 1 July 2009. Also, Faculty and Former Faculty and Senior Leader Interviews and Communications, AU and USAFA, April – November 2009.

[738] Faculty Panels, HASC Staff Visits, AU and USAFA, 8-9 April 2009 and 28-29 June 2009, respectively. Also, USAFA Faculty and Former Faculty and Senior Leaders, HASC Staff Interviews and Correspondence, April – December 2009.

[739] *Air University Instruction 36-2308, Academic Freedom*, 22 August 2008. See also MajGen Forsyth, Response to Question for the Record, HASC O&I Subcommittee Hearing, *Thinkers and Practitioners*, CHARRTS No.: HASCOI-05-005.

[740] Faculty Panels, HASC Staff Visit, AU, 8-9 April 2009. See also Jeffrey Record, "The Creeping Irrelevance of U.S. Force Planning," Strategic Studies Institute, 19 May 1998 and "Bounding the Global War on Terrorism," Strategic Studies Institute, December 2003.

[741] RADM Wisecup, HASC Response to Questions for the Record, HASC O&I Subcommittee Hearing, *Raising Thinking*, CHARRTS No.: HASCOI-06-014, and HASC O&I Subcommittee Hearing *Thinkers and Practitioners*, Response to Questions for the Record, CHARRTS No.: HASCOI-05-005. He also stated, "Faculty members are allowed great scope for experimenting with different teaching methods and for expressing different points of view in the classroom."

[742] ACSC Staff, Response to Question for the Record, HASC O&I Subcommittee Hearing, *Raising Thinking*, CHARRTS No.: HASCOI-06-014.

[743] USAFA Faculty Panel, HASC Staff Visit, 29 June 2009.

[744] USAFA, *FOI 35-101*, *Clearance of Material for Public Release and Academic Freedom*, 24 January 2008.

[745] Faculty Panel and Discussion with BrigGen Born, HASC Staff Visit, USAFA, 29 June 2009. Faculty and Former USAFA Faculty, HASC Staff Interviews, April – December 2009.

[746] USAFA, *FOI 35-101*, 3 and 6-8. Also Faculty Panel and Discussion with BrigGen Born, HASC Staff Visit, USAFA, 29 June 2009. Also Faculty and Former USAFA Faculty, HASC Staff Interviews, April – December 2009.

[747] The Higher Learning Commission, "Assurance Section" and "Advancement Section," *Report of A Comprehensive Evaluation Visit to United States Air Force Academy*, 30 and 7, respectively. Senior Leader and Faculty Panels, HASC Staff Visit, USAFA, 29 June 2010. See USAFA, *FOI 35-101*, *Clearance of Material for Public Release and Academic Freedom*, 24 January 2008, 1-3.

[748] Faculty Panel, HASC Staff Visit, USAFA, 28-29 June 2009.

[749] The Higher Learning Commission, *A Commission of the North Central Association of Colleges and Schools*, "Assurance Section" and "Advancement Section," *Report of A Comprehensive Evaluation Visit to United States Air Force Academy*, USAF Academy, CO, 27-29 April 2009, 30 and 7 respectively. Also USAFA Faculty Panel, HASC Staff Visit, 28-29 June 2009.

[750] Dr. Fran Pilch email, "Re: PME and Academic Freedom," USAFA, 2 December 2009.

[751] USMA Faculty Panel, HASC Staff Visit, 30 June – 1 July 2009. USMA, *Academic Freedom Advisory Committee Report*, 22 May 2009.

[752] BG Patrick Finnegan, email to HASC Staff, 23 October 2009. See also, USMA, *Institutional Self Study Report*, June 2009, 32-45, and Middle States Commission on Higher Education, "Final Report to the Faculty, Administration, Trustees, Cadets of the United States Military Academy," 13-16 September 2010.

[753] Brian T. Yeh, "Copyright Issues and Department of Defense Personnel," Congressional Research Service, 25 June 2009.

[754] *Ibid*. Also, CRS Briefing to HASC Staff, 23 September 2009. CRS Meeting with HASC O&I Subcommittee Chairman, 25 September 2009. CRS Briefing to HASC O&I Subcommittee Chairman October 2009. Request from Representative Todd R. Platts, 17 November 2009. See also 17 U.S.C. §§105 and 101. See also Bonnie Klein, "*The Office of Management and Budget Circular OMB-A 130* established the policy for the Management of Federal Government Information Resources," Vol. 17, No. 4, ISSN 1070-4795, 34. 5 C.F.R.

§2635.807. *USMA Faculty Handbook*, West Point, NY, December 2005, 17-18. MCU, *Title 10 Faculty Handbook*, Quantico, VA, June 2008. CGSC, *Title 10 Civilian Faculty Manual CGSC Pamphlet 690-1*, August 2008. BG Cardon, Response to Question for the Record, HASC O&I Subcommittee Hearing, *Raising Thinking*, CHARRTS No.: HASCOI-06-020. Faculty Panel, Staff Visit, National, 3 April 2009, and Faculty Interviews (including Harvey Rishikof, Esq., 25 June 2009). Also Faculty Panels, Staff Visits, USAWC, NWC, and AU, 5-9 April 2009. Also, Faculty Panels, HASC Staff Visits, USAFA and West Point, 28 June – 1 July 2009.

[755] Faculty Panels, HASC Staff Visits, USAWC, NWC, and AU, 6-9 April 2009. Also Faculty Panels, HASC Staff Visits, USAFA and USMA, 28 June – 1 July 2009. Request from Representative Todd R. Platts, 17 November 2009. See also 17 U.S.C. §§105 and 101.

[756] 17 U.S.C. §§105 and 101. See also Bonnie Klein, "*The Office of Management and Budget Circular OMB-A 130* established the policy for the Management of Federal Government Information Resources," Vol. 17, No. 4, ISSN 1070-4795, 34.

[757] Yeh, "Copyright Issues." See 17 U.S.C. § 106 and 501. Also, CRS Briefing to HASC Staff, June 2009, and CRS Briefing to HASC, O&I Members, 17 October 2009. See also faculty handbooks.

[758] 5 C.F.R. §2635.807.

[759] Cardon, Response to Question for the Record, HASC O&I Subcommittee Hearing, *Raising Thinking*, CHARRTS No.: HASCOI-06-020.

[760] National Faculty Panel, HASC Staff Visit, 1-3 April 2009, and National Faculty Interviews, HASC Staff, March – December 2009. Also Faculty Panels, HASC Staff Visit, USAWC, NWC, and AU, 6-9 April 2009.

[761] H.R. 962, 109th Congress, 1st Session, 17 February 2005. This bill would have amended Chapter 53 of Title 10 by inserting section 1033a for the Department of Defense, Chapter 9 of Title 14 by inserting section 198 for the Coast Guard, and Title XIII of the Merchant Marine Act of 1936 (46 App. U.S.C. 1295 et seq.) by adding section 1308.

[762] Yeh, "Possible Approaches to Amending the Copyright Law to Allow faculty of Military Service Academies and Department of Defense professional Schools to Secure Copyrights For Their Scholarly Works," 6 November 2009, 2.

[763] *Ibid.*, 2-5.

[764] See for example, Chairman of the Joint Chiefs of Staff, *Process for Accreditation of Joint Education (PAJE) College of Naval Warfare*, 14-18 May 2007, updated 4-7 May 2009.

[765] Senior Leader and Faculty Panels, CODELs Snyder, ICAF and National, 31 March and 29 April 2009, respectively. Senior Leader and Faculty Panels, HASC Staff Visits, National 3 April 2009; USAWC, NWC, and AU, 6-9 April 2009; JFSC 30 April – 1 May 2009; and, USAFA, 29-30 June 2009. Senior Leader and Faculty Interviews, HASC Staff, April – December 2009.

[766] *Ibid.*

[767] ICAF and NWC Faculty Briefings, HASC Staff, 2 June 2009 and 9 June 2009 and Faculty Interviews, March – November 2009.

[768] Larson, 2-9 and 2-10. Air Force Instruction 36-3501, Air *Force Academy Operations*, 28 April 2008, para. 2.13.16.3. See also Born, Response to Question for the Record, HASC O&I Subcommittee Hearing, *Beyond Service Core Competency*, CHARRTS No.: HASCOI-08-037.

[769] Larson, ES-3, 2-11 and 2-12.

[770] See for example, Chairman of the Joint Chiefs of Staff, *Process for Accreditation of Joint Education (PAJE) College of Naval Warfare*, 14-18 May 2007, updated 4-7 May 2009.

[771] Judith Hicks Stiehm, *The U.S. Army War College: Military Education in a Democracy*, (Temple University Press, Philadelphia, PA, 2002), 70-77.

[772] National Faculty Panel, CODEL Snyder, 29 April 2009. Murray, Written Testimony, HASC O&I Subcommittee Hearing, *Charting the Course*, 23.

[773] Senior Leader, Faculty, and Student Panels, CODELs Snyder, ICAF and National, 13 March and 29 April 2009, respectively. Senior Leader, Faculty, and Student Panels, HASC Staff Visits, USAWC, NWC, and AU, 6-9 April 2009, and JFSC, 30 April – 1 May 2009. Murray, Written Testimony, HASC O&I Subcommittee Hearing, *Charting the Course*, 16.

[774] CJCS, *Process for Accreditation of Joint Education (PAJE), U.S. Marine Corps Command and Staff College*, Quantico, VA, 22-26 September 2008, 14, see also 4.

[775] CJCS, *Process for Accreditation of Joint Education (PAJE), U.S. Air War College*, Maxwell Air Force Base Alabama, 20-23 October 2008, 14.

[776] CJCS, *Process for Accreditation of Joint Education (PAJE), Marine Corps War College*, Marine Base Quantico, VA, 12-14 January 2009, 5.

[777] CJCS, *Process for Accreditation of Joint Education (PAJE), National War College, Fort Leslie J. McNair*, Washington, D.C., *14- 19 November 2009*, 18.

[778] *Ibid., 18-19.*

[779] Faculty Panels, CODEL Snyder 29 April 2009, and HASC Staff Visit, National, 3 April 2009.

[780] Faculty Panels, CODELs Snyder ICAF and National, 31 March and 29 April 2009, respectively. National Faculty Panel, HASC Staff Visit, National, 3 April 2009. Also Faculty Interviews, HASC Staff, March – August 2009.

[781] CJCS, *Process for Accreditation of Joint Education (PAJE), Marine Corps War College*, Marine Base Quantico, VA, 12-14 January 2009, 7.

[782] *The Skelton Report*, 133.

[783] *Ibid.*, 150-155.

[784] *Ibid.*, 133 and 150-158. And, Faculty and Student Panels, CODELs Snyder, ICAF and National, 31 March, 29 April 2009, and 4 June 2009, respectively; MCU 20 May 2009. Also Faculty and Student Panels, HASC Staff Visits, MCU, 27 March 2009; National, 3 April 2009; USAWC, NWC, and AU, 6-9 April 2009; JFSC, 30 April – 1 May 2009; and, CGSC, 26-29 May 2009.

[785] USAF Briefing, HASC Staff, 26 March 2009.

[786] MCU Briefing, HASC Staff, 23 March 2009.

[787] CGSC Faculty and Student Panels, HASC Staff Visit, 26-27 May 2009. Also LTG David Ohle (Ret.), HASC Staff Interview, 31 July 2009 and January – March 2010.

[788] Senior Leader and Faculty Panels, HASC Staff Visits, USAWC and CGSC, 6-7 April and 26-27 May 2009, respectively. Also, Army Briefing, HASC Staff, 26 March 2009. General George Casey, Member Meeting, 21 October 2009. General Martin Dempsey and LTG David Barno (retired), HASC Staff Interviews, April – December 2009.

[789] Navy Briefing and Navy (N15) Briefings to HASC Staff, 31 March and 11 May 2009, respectively.

[790] RADM Wisecup, Written Testimony, HASC O&I Subcommittee Hearing, *Raising Thinking*, 2. Also, Navy Briefing and Navy (N15) Briefing to HASC Staff, 31 March and 11 May 2009, respectively. Also Lutterloh, Testimony, HASC O&I Subcommittee Hearing, *Investing in Our Military Leaders.*

[791] Navy (N15) Briefing to HASC Staff, 11 May 2009. Also Lutterloh, HASC O&I Hearing, *Investing in Our Military Leaders*, Testimony.

[792] Faculty and Student Panels, CODELs Snyder, ICAF and National, 31 March and 29 April 2009, respectively. Student Panel, Member Meeting, 5 June 2009. Also Faculty and Student Panels, HASC Staff Visits, National 3 April 2009; USAWC, NWC, and AU, 6-9 April; and, CGCS, 26-27 May 2009.

[793] *Ibid.*

[794] National, Faculty Panel, CODEL Snyder, 29 April 2009. Student Panel, HASC O&I Member Meeting, 5 June 2009. Faculty and Student Panels, HASC Staff Visit to National, 3 April 2009. Also Faculty and Student Panels, HASC Staff Visits, USAWC, NWC, and AU, 6-9 April 2009.

[795] National Faculty Panel, CODEL Snyder, 29 April 2009. National Student Panel, HASC O&I Member Meeting, 5 June 2009. National Faculty and Student Panels, HASC Staff Visit, 3 April 2009. Faculty and Student Panels, HASC Staff Visits, USAWC, NWC, and AU, 6-9 April 2009. Faculty Interviews, HASC Staff, March – December 2009. Kohn, Testimony and Written Testimony, HASC O&I Subcommittee Hearing, *Another Crossroads*, 2. John Williams, Testimony and Written Testimony, HASC O&I Subcommittee Hearing, *Charting the Course*, 1.

[796] *Ibid.*

[797] Lutterloh, Responses to Questions for the Record, HASC O&I Subcommittee Hearing, *Investing in Our Military Leaders.*

[798] *The Skelton Report*, 151-154, and *OPMEP*, Enclosure B.

[799] CGSC Senior Leader Panel, HASC Staff Visit, 26-27 May 2009.

[800] Caldwell, Sitterly, Lutterloh, and Spiese, Response to Questions for the Record, HASC O&I Subcommittee Hearing, *Investing in Our Military Leaders*, CHARRTS No.: HASCOI-07-016, HASCOI-07-026, HASCOI-07-035, and HASCOI-07- 044, respectively.

[801] HASC Staff Visit to USAWC, NWC, and AU, 5-9 April 2009.

[802] Senior Leader, Faculty, and Student Panels, CODELs Snyder, ICAF and National, 31 March 2009 and 29 April 2009, respectively. Senior Leader and Faculty Panels, HASC Staff Visits, National, 3 April, and USAWC, NWC, and AU, 5-9 April 2009. Also ICAF Senior Leader and Faculty Interviews, HASC Staff, 2 June and 29 July 2009.

[803] LTG Caldwell, Response to Questions for the Record, HASC O&I Subcommittee Hearing, *Investing in Our Military Leaders*, CHARRTS No.: HASCOI-07-015.

[804] *Ibid.*

[805] Sitterly, Response to Questions for the Record, HASC O&I Subcommittee Hearing, *Investing in Our Military Leaders*, CHARRTS No.: HASCOI-07-043.

[806] Senior Leader and Faculty Panels, HASC Staff Visits, USAWC, NWC, and AU, 5-9 April 2009 and ICAF Senior Leader and Faculty Interviews, HASC Staff, 2 June and 29 July 2009.

[807] Steve Harris, DSLDP Program Manager, Civilian Personnel Management Service, Briefing Slides, October 2009.

[808] NDAA FY 2010, P.L. 111-84 Sec. 1112.

[809] Senior Leader, Faculty Panels, and Student Panels, HASC Staff Visit, MCU, 27 March 2009, and Senior Leader, Faculty Panels, and Student Panels, HASC Staff Visits, USAWC, NWC, and AU, 5-9 April 2009.

[810] CGSC Senior Leader Panel, HASC Staff Visit, 26-27 May 2009.

[811] LTG Caldwell, HASC Staff Interview, CGSC, 26-27 May 2009.

[812] Gen Norton Schwartz, Chief of Staff of the U.S. Air Force, HASC O&I Member Meeting, 1 October 2009.

[813] Senior Leader, Faculty, and Student Panels, HASC Staff Visits, USAWC, NWC, and AU, 5-9 April 2009; JFSC, 30 April – 1 May 2009; and CGSC, 26-27 May 2009.

[814] Senior Leader, Faculty, and Student Panels, CODELs Snyder, ICAF, National, and MCU, 31 March, 29 April, and 20 May 2009, respectively. Senior Leader and Student Meetings, HASC O&I Members, 9 and 4 June 2009, respectively. Senior Leader, Faculty, and Student Panels, HASC Staff Visit, MCU, 27 March 2009. Senior Leader, Faculty, and Student Panels, HASC Staff Visits, USAWC, NWC, and AU, 5-9 April 2009.

[815] Senior Leader, Faculty, and Student Panels, HASC Staff Visit, National, 3 April 2009. Senior Leader, Faculty, and Student Panels, HASC Staff Visits, USAWC, NWC, and AU, 6-9 April 2009. Each school uses a different pre-selection English assessment for international students. Depending on their English competency (usually measured by the TOEFL (Test of English as a Foreign Language) tool), some may compete for the PME schools' or nearby universities' master's degrees.

[816] JCS/DJ-7, Briefing to HASC Staff, 17 March 2009. Senior Leader, Faculty, and Student Panels, HASC Staff Visit, National, 3 April 2009. Senior Leader, Faculty, and Student Panels, HASC Staff Visits, USAWC, NWC, and AU, 6-9 April 2009.

[817] John Charlton, NDU International Programs Briefing to HASC Staff, 22 May 2209. VADM Ann Rondeau, HASC O&I Member Meeting, 23 July 2009. See also, ADM Mullen speech at NDU for Wilson-Rondeau change of presidency and direction to NDU, 10 July 2009.

[818] Charlton; Rondeau, 23 July 2009; International Programs Discussion, HASC Staff Visit, NWC, 7-8 April 2009; and, JCS/DJ-7, Briefing to HASC Staff, 17 March 2009.

[819] ICAF Senior Leader, Faculty, and Student Panels, CODEL Snyder, 31 March 2009. NWC Senior Staff, Faculty, and Student Panels, HASC Staff Visit, 7-8 April 2009. Also, Charlton.

[820] Charlton, and ICAF Senior Leader, Faculty, and Student Panels, CODEL Snyder, 31 March 2009. National Faculty Panel, CODEL Snyder, 29 April 2009. Senior Leader and Student Meetings, HASC O&I Members, 9 and 4 June 2009, respectively. National Senior Leader, Faculty, and Student Panels, HASC Staff Visit, 3 April 2009. Also, Senior Leader, Faculty, and Student Panels, HASC Staff Visits, USAWC, NWC, and AU, 5-9 April.

[821] USAWC Faculty Panel, HASC Staff Visit, 6-7 April 2009.

[822] NWC Faculty Panel, HASC Staff Visit, 7-8 April 2009.

[823] Faculty and Student Panels, CODELs Snyder, ICAF and National, 31 March and 29 April 2009, respectively. Faculty and Student Panels, HASC Staff Visits, National and NWC, 3 April and 7-8 April 2009, respectively. Also, Charlton.

[824] 10 U.S.C. §2167(a). See also NDAA FY 2010, *P.L. 111-84*, Sec. 532.

[825] Senior Leaders, Faculty, and Student Panels, CODEL Snyder, ICAF, 31 March 2009.

[826] *The Skelton Report*, 156.

[827] *Ibid., 156-157.*

[828] Under Secretary of Defense (Comptroller) Robert Hale, Letter to O&I Subcommittee Chairman Vic Snyder, 5 January 2010.

[829] *Ibid.*

[830] OPMEP, E-2.

[831] DDJS-ME Memorandum, *Process for Accreditation of Joint Education (PAJE), College of Naval Command and Staff*, 25-30 October 2009, 11-12.

[832] DDJS-ME Memorandum, *Process for Accreditation of Joint Education (PAJE), U.S. Army War College*, 21-25 September 2009, 9-10.

[833] DDJS-ME Memorandum, *Process for Accreditation of Joint Education (PAJE), U.S. Marine Corps Command and Staff College*, 22-26 September 2008, 8-9.

[834] DDJS-ME Memorandum, *Process for Accreditation of Joint Education (PAJE), College of Naval Command and Staff*, 25-29 February 2008, 7-8.

[835] CODELs Snyder, MCU, 20 May 2009 and 4 March 2010.

[836] *Ibid.*, and Col Brian D Beaudreault, Testimony, HASC O&I Subcommittee Hearing, *Beyond Service Core Competency.*

[837] CODELs Snyder to ICAF, 31 March 2009, and National War College, 29 April 2009. HASC Staff Visit, National, 3 April 2009. Also HASC Staff Interviews at ICAF, National, and National Defense University, March-May 2009. See also, Breslin-Smith, Testimony, HASC O&I Subcommittee Hearing, *Another* Crossroad? 20 May 2009.

[838] Dr. John Jay Williams, Testimony, Congress, U.S. House of Representatives, Committee on Armed Services, Hearing before the Subcommittee on Oversight and Investigations, *Charting the Course for Effective Professional Military Education*, 10 September 2009.

839 CODEL Snyder, Marine Corps University, 4 March 2010.

840 *Ibid.*

841 William A. Navas, Jr., "Working Paper: National Security Professional Development Program," 20 March 2009.

842 *Ibid.*

843 Dr. Janet Breslin-Smith, Written Testimony, Congress, U.S. House of Representatives, Committee on Armed Services, Hearing before the Subcommittee on Oversight and Investigations, *Another Crossroads? Professional Military Education Twenty Years after the Goldwater-Nichols Act and the Skelton Panel*, 20 May 2009, 6.

844 Dr. James Jay Carafano, Written Testimony, HASC O&I Subcommittee Hearing, *Another Crossroads?* 20 May 2009, 2-5.

845 USNAINST 1610.3G, 21 December 2005, 30.

846 Richard Cheney and Bill Taylor, *Professional Military Education: An Asset for Peace and Progress: A Report of the CSIS Study Group on Professional Military Education*, The Center for Strategic and International Studies, Washington D.C., March 1997.

847 CJCSI 1800.01D, *Officer Professional Military Education Policy (OPMEP)*, 15 July 2009, E-D-3 and E-E-3.

848 LtGen John M. Paxton, Response to Question for the Record, Congress, U.S. House of Representatives, Committee on Armed Services, Hearing before the Subcommittee on Oversight and Investigations, *Investing in Our Military Leaders: The Role of Professional Military Education in Officer Development*, 28 July 2009, CHARRTS No.: HASCOI-07-003. See also LtGen Paxton cite from Joint Publication 1, *Doctrine of the Armed Forces of the United States*, 2 May 2007, w/Change, 20 March 2009, and General Peter Pace, "CJCS Vision for Joint Officer Development," November 2005, 1. Also Chairman of the Joint Chiefs of Staff, *Process for Accreditation of Joint Education (PAJE), United States Naval War College, College of Naval Command and Staff (Resident)*, Newport, Rhode Island, 25-30 October 2009, 18.

849 Cheney and Taylor, 47.

850 *OPMEP*, A-A-6.

851 LtGen Paxton, Response to Question for the Record, HASC O&I Subcommittee Hearing, *Investing in Our Military Leaders*, 28 July 2009, CHARRTS No.: HASCOI-07-003. See also cite from *OPMEP*, 15 July 2009, A-A-1, A-A-3, E-E-3, E-F-4, E-G-4, E-I-3, and E-K-2.

852 MajGen Robert P. Steel and RADL Garry E. Hall, Response to Question for the Record, Congress, U.S. House of Representatives, Committee on Armed Services, Hearing before the Subcommittee on Oversight and Investigations, *Thinkers and Practitioners: Do Senior Military Education Schools Produce Strategists?* 4 June 2009, CHARRTS No.: HASC-05-012. See also, LTG William B. Caldwell IV, Dan Sitterly, MajGen (S) Melvin Spiese, and Scott Lutterloh, Responses to Question for the Record for HASC O&I Hearing, *Investing in Our Military Leaders*, CHARRTS No.s: HASCOI-07-012, - 040, -31, and -022, respectively. RADM James P. Wisecup, MG Robert M. Williams, MajGen Maurice "Maury" Forsyth, and Col Michael Belcher, Responses to Question for the Record for HASC O&I Subcommittee Hearing, *Thinkers and Practitioners*, 4 June 2009, CHARRTS No.: HASC-05-012.

853 This was the consensus among the recent literature and in interviews with current chairs of ethics at the PME institutions.

854 Cheney and Taylor, 47-48.

855 *Ibid.*, 49.

856 *Ibid.*, 50.

857 West Point established the Simon Center for the Professional Military Ethic in 2000. The Naval Academy renamed their Center for the Study of Professional Military Ethics established in 1998, the Vice Admiral James B. Stockdale Center for Ethical Leadership in 2006. The Air Force Academy established the Center for Character Development in 1993. The Marine Corps University Lejeune Leadership Institute has an Ethics Branch, and the Command and General Staff College received a $6.1 million donation from H. Ross Perot in November 2009 to establish an interagency center and the new General Hugh Shelton Chair in Ethics.

858 Interviews with Director Dr. Albert C. Pierce, 28 September 2009, and General (retired) Richard B. Myers, 26 January 2010.

859 Admiral Michael G. Mullen, Interview for *Joint Forces Quarterly*, Issue 54, 3d Quarter, 2009, 7-11.

In: Professional Military Education
Editor: Michelle P. Jacobs
ISBN: 978-1-62948-814-1

Chapter 2

STATEMENT OF BRIGADIER GENERAL SEAN B. MACFARLAND, DEPUTY COMMANDANT, ARMY COMMAND AND GENERAL STAFF COLLEGE. HEARING ON "CONTINUED ENGAGEMENT: DOD RESPONSES TO THE HASC REPORT ON PROFESSIONAL MILITARY EDUCATION"*

INTRODUCTION

Chairman Snyder, Representative Wittman, Honorable Members of the Subcommittee, I am BG Sean MacFarland, Deputy Commanding General of the Combined Arms Center. On behalf of the Chief of Staff Army, I would like to thank you for the opportunity to speak with you today about the present status of Professional Military Education in the Army and discuss what we are doing to better prepare leaders to operate with competence and character in the future.

The Army concurs with the Subcommittee's conclusion, the Professional Military Education (PME) system is basically sound, and agrees there are "systemic" and "institutional" areas requiring our continued focus. We recognize the Subcommittee's report provides valuable insights to assist us in our continual assessment of PME and efforts to provide the best educational opportunities to our Soldiers. My comments focus on three areas of interest: Framework for Leader Development, Building a Strategic Bench, and Army PME Institutions.

FRAMEWORK FOR LEADER DEVELOPMENT

First, the Subcommittee's report recommended the Services review their officer development timelines from a holistic perspective to explore innovative avenues to develop

* This is an edited, reformatted and augmented version of a Statement, presented November 30, 2010 before the House Armed Services Committee, Oversight and Investigations Subcommittee.

their respective officer corps through education, training, and experience. The Army supports this recommendation. We fully recognize Professional Military Education is paramount to an officer's career development. We continuously review our officer Professional Military Education to ensure it remains relevant to our force and national needs.

The strategic environment is growing more complex, increasing the demand on education within innovative and dynamic leader development. The changing environment, nine years of war and the requirement to sustain our readiness causes us to think fundamentally different about the integration of Education, Training, and Experiences for our leaders. We are working diligently to adapt our institutions and policies to achieve better balance of Professional Military Education within leader development. This is particularly evident in our PME approach, as we must account for educational changes in our leader development and personnel management policies. We are driving these changes through development of a comprehensive Army Leader Development Strategy (ALDS) focused on educating the next generation of Army leaders.

The Army's Leader Development Strategy seeks to develop agile, adaptive, and innovative leaders for our Army within a flexible, relevant, and enduring framework of training, education, and experience. In an uncertain and complex future security environment Army leader development must prepare leaders to operate with competence, confidence, and character in ambiguous, frequently changing circumstances. Our leaders deserve the best possible leader development process to enable them to effectively lead our Soldiers and protect and promote our national interests and sustain the public trust. In this era of persistent conflict, we must increase our efforts to develop each of our leaders, and we must ensure that we are preparing our most talented leaders to lead our Army into the future.

Professional Military Education is essential to an officer's career development. We are shaping the future of our PME system to progressively develop leaders to prevail in the Full Spectrum Operations security environment of the 21st Century. Our Professional Military Education continues to reflect our adaptability to changing conditions and demonstrates the strong emphasis our Army Senior Leaders place on education through agility, adaptability, innovation, and versatility.

Over the past several years, as many recognize, we have become an Army out of balance. Less well known, but increasingly evident, is the fact we are out of balance in developing our leaders for the demands of Full Spectrum Operations. Overemphasizing operational experience at the expense of Professional Military Education and assignment to broadening experiences has contributed to the imbalance.

The Army recognizes it must restore this balance, while simultaneously setting conditions for future success. Our future readiness demands we continue to modernize, adapt our institutions, and transform Soldier and leader development to ensure our campaign capable force is versatile, expeditionary, agile, lethal, sustainable, and interoperable. Adapting our institutions increases efficiency and effectiveness in providing trained and ready forces for combatant commanders. We appreciate your frank evaluation of our schools and educational processes.

Transforming how we train Soldiers and develop agile and adaptive leaders, based on the lessons we continue to learn in Afghanistan and Iraq, is paramount to the success of Full Spectrum Operations in dynamic and complex operating environments.

The Army's Training and Doctrine Command is leading our Human Capital Enterprise. This new initiative will allow synchronization of effort for Army leader development and personnel policies. An important part of the Human Capital Enterprise mission is to make a concerted effort to restore the balance between the pillars of education, training, and experience.

[Note: References in the report depict the Human Capital Enterprise in a manner implying that this organizational entity is part of the Army's approved structure. At this time, the Human Capital Enterprise is one of four Core Enterprises, which are functioning on a trial basis as part of the Army's transformation efforts. The concept of Core Enterprises has not yet been decided; nor have they been approved or incorporated into the Army's official structure. If approved, the current intent is for the appropriate Assistant Secretary of the Army to lead their functional Core Enterprise with support from the Army Staff and Army Commands.]

BUILDING A STRATEGIC BENCH

Second, the Subcommittee's report recommended the Services carefully review and further develop relevant processes for identifying and cultivating strategists. The Army's challenge is to grow adaptive innovative leaders who learn faster, understand better, and successfully handle uncertainty. Our leaders must be able to effectively work with our Joint, Interagency, Intergovernmental, and Multinational teammates. We must continually challenge our leaders with a competitive learning environment, and it is in this environment where we must prevail. PME alone is no longer adequate. We have to recognize the officers with high potential and provide them with challenging experiences. To do this, the Army offers multiple broadening opportunities outside the realm of our traditional PME to educate our officers throughout their career timeline. These experiences are designed to develop a wider range of knowledge and skills; augment understanding of the spectrum of military missions; promote application of specialized skills; and expand awareness of other agencies, organizations, and environments. I would like to highlight a few of these broadening programs.

Through the Advance Civil Schooling (ACS) Program, the Army resources up to 412 fully funded resident degree programs placing officers into positions requiring a graduate or doctoral level degree. The ACS program also contributes to an inventory of officers who possess requisite academic credentials and technical skills to meet existing force structure requirements, as well as meeting future capabilities envisaged by the Army Educational Requirements System (AERS). Additionally, ACS educates officers about emerging technologies, who can then transform relevant information from the laboratory to real world applications on the battlefield.

Since 2006 we have offered the Expanded Graduate School Program (EGSP) for up to 600 officers per year. We identify officers prior to commissioning who we believe have the potential for long and productive service. At the 10-12 year mark in their career, selected officers have the opportunity to attend a US civilian graduate school and are encouraged to study an approved discipline to enhance the competencies required in an expeditionary Army--emphasizing cultural awareness, regional knowledge, foreign language, governance, diplomacy, national security, or social sciences to reinforce operational skills.

The Joint Chiefs of Staff/Office of the Secretary of Defense (JCS/OSD) Internship is another great opportunity for our young captains with outstanding potential for service to experience the Joint environment early in their career. They complete a three-phase, degree-producing program. Completion of Phase One results in a Master's Degree in Policy Management from Georgetown University's Public Policy Institute. Phase Two is a one-year internship within a single office in the Joint Staff or the Office of the Secretary of Defense. Phase Three is a one-year internship on the Army Staff. Assignments to Army Staff positions are based upon unique skills acquired during both of the previous phases of the program. Like the Joint Staff and OSD Staff, the Army Staff places interns into demanding roles as staff officers, to leverage their knowledge base to represent the Army within the Pentagon and the Department of Defense.

The Army Congressional Fellowship Program (ACFP) educates selected Army captains and civilians in all aspects of congressional activities, emphasizing those matters regarding the Department of Defense (DOD). A three-year program, ACFP includes a master's degree in legislative affairs at George Washington University, service on the staff of a Member of Congress, and utilization on the Army or Joint Staff in a legislative liaison duty position. The program provides insights into the dimensions and complexities of congressional responsibilities and their relationship to the total process of government. In the past year we have increased ACFP opportunities from 12 to 24.

The Functional Area (FA) 59, Strategist career field is another component for building our strategic bench. The Army selects most strategists between year seven and 14 of commissioned service and educates them with Advanced Civilian Schooling, Intermediate Level Education, the on-line Defense Strategy Course, and the resident Basic Strategic Arts Program. FA59 utilizes the School of Advanced Military Studies and Civilian Fellowships to develop and strengthen core competencies. After these educational opportunities the strategists serve in a variety of headquarters from division to combatant command level. Senior FA59s (COL-BG) will then work within the Interagency, Office of the Secretary of Defense, Army and the Joint Staff. In these positions, they participate in campaign planning, policy development and work with interagency partners on a wide variety of critical national security issues. A concern raised by the Committee was the lack of viability in promotion prospects for senior FA59s. This trend is shifting as the second FA59 officer has recently been nominated for promotion to General Officer.

Established in 2009, the Command and General Staff College (CGSC) Interagency Fellowship Program is another unique opportunity to help build our strategic bench. This program offers Army majors a greater understanding of the role of our interagency partners. They have the opportunity to gain personal knowledge about the capabilities, missions, procedures, and requirements of Federal departments, agencies, and Non-Governmental Organizations (NGOs) through participation in a 9-12 month fellowship. We currently have 28 majors working with 13 agencies. This program has also allowed interagency organizations to send their employees to CGSC, thus enhancing Intermediate Level Education (ILE).

The Advanced Military Studies Program (AMSP) is another critical component of our efforts to develop officers as operational planners. Majors may apply while attending Intermediate Level Education or may be nominated from field units after having completed their Intermediate Level Education course. Each candidate undergoes a rigorous selection process. AMSP educates leaders to be critical thinkers and operational planners by teaching officers to understand the theory of operational art, to define security problems using design methodology, to understand and evaluate the strategic context for operational art, to understand the evolution of operational art through history, and to be competent at preparing military plans using both the Joint Operation Planning Process and the Army Military Decision Making Process. The Army recognized the growing need for such planners in 2008, and increased the student population from 68 to 108 per year.

The Command and General Staff College also offers a senior level program for officers in the rank of LTC-COL. The Advanced Operational Art Studies Fellowship (AOASF) focuses on the art of command and leadership for national and theater– strategic strategy development and campaign planning. AOASF seeks to educate the future leaders of our Armed Forces, our Allies, and the Inter-Agency at the graduate level to be agile and adaptive leaders who think critically at the strategic level to solve complex, ambiguous problems.

On behalf of the Chief of Staff of the Army, the Army War College administers a mature and competitive Senior Service Fellowship program designed to broaden up to 75 senior officers in a wide variety of disciplines at 35 prestigious educational, governmental, and non-governmental institutions. This program provides the Army with a cohort of senior leaders possessing a broad range of competencies and experiences to better meet the demands of the strategic and institutional environments of the 21st Century.

Other broadening opportunities for senior leaders include fellowships at the Brookings Institute; the Council of Foreign Relations; and the Harvard University, Weatherhead Center for International Affairs. Additionally, there are Regional Fellowships, including the Asia-Pacific Center of Security Studies; the George C. Marshall Center; and the State Department Planning and Policy Fellowship. Other Fellowships are conducted at the White House, Congress, Arroyo Center, and RAND.

The focal point of Army General Officer Training is the Army Strategic Leadership Development Program (ASLDP). This program supports Army General Officer development by requiring attendance at the Basic, Intermediate, Advanced and Senior courses as officers are promoted from Brigadier to Lieutenant General. Other courses cover Army Senior Leader Communication, Force Integration, Battle Command, Legal, Senior Commander, National Security Studies, National and International Security, Defense Policy, Senior Joint Information Operations, National Security Decision Making, Language and Cultural Awareness and Individual Terrorism Awareness.

The Army War College also offers several short courses for general/flag officers including: Combined/Joint Force Land Component Commander Course (C/JFLCC), Adjutant's General National Security Seminar (AGNSS), and Senior Reserve Component Officers' Course (SRCOC). The Army War College also supports the Joint Flag Officer Warfighting Course (JFOWC) hosted by Air University.

Professional Military Education Institutions

Finally, the quality of our Professional Military Education Institutions is important to us as it directly influences the education our Soldiers receive. The Subcommittee's report referenced a number of valid concerns I would like to address.

The Subcommittee's report referenced a continuing debate concerning how the school presidents, commanders, commandants, and directors should be selected and how long they should remain in their positions. The Army takes very seriously both the professional and academic credentials of its PME leaders. The Army has specific requirements for the leadership positions at both CGSC and US Army War College (USAWC). These requirements are codified into the manning documents at the Human Resources Command.

The Commandant, USAWC should possess an extensive background in both professional and academic domains. The professional background should include extensive experience at the theater and strategic levels. Academically, Commandants should possess, at a minimum, a masters degree and preferably a terminal degree in a discipline appropriate to Senior Level College curriculum. Prior teaching experience, especially at the graduate level or within the JPME educational system is a decided plus. As befits a professional development school, leadership capabilities of the Commandant should be a paramount consideration in selection.

The ideal candidate for the Dean of Academics, USAWC would be an Army colonel who has commanded at the battalion level or higher, has extensive professional experience at the operational or strategic level, is a Senior Level College graduate and possesses a terminal degree in a discipline relevant to the Army War College academic program. Alternatively, a civilian candidate should have a terminal degree in an appropriate discipline and military or civilian experience at the strategic level. In either case, teaching experience, particularly at the graduate level or within the JPME system is highly desired. Because ideal conditions cannot always be met and because war colleges are professional development institutions, preference should be given to professional experiences. Department chairs should have qualifications similar to the Dean of Academics.

At CGSC, school directors must be a colonel or civilian equivalent, be graduates of senior level colleges, hold masters degrees, have leadership experience in the operational environment, and have commanded battalions or higher. Their civilian counterparts must have terminal degrees, be considered subject matter experts in their chosen area of expertise, have leadership experience, and have a thorough understanding of Army structure and organization. The civilian Dean of Academics at CGSC holds a terminal degree and has the requisite experience and background commensurate with the position. Having a terminal degree is extremely important for the Dean of Academics because he is responsible for faculty and curriculum. In addition, he provides subject matter expertise to other institutions that comprise the Army's Leader Development and Education (LD&E) enterprise. Under the Dean's purview, he is responsible for quality assurance, accreditation, faculty development, registrar functions, graduate degree programs, the library, the captains career course, academic chairs, digital leader development center, and academic operations.

The quality of the faculty at our PME institutions is extremely important to us. The report raised a concern about the hiring and retaining of Title 10 civilian professors. We have found hiring under Title 10 to be a real strength. The law offers the ability to identify specific faculty requirements, provides flexibility in hiring searches, and allows for competitive salary and benefits. Since all faculty are on term appointments requiring review at regular intervals, the Title 10 process allows CGSC and USAWC greater flexibility to retain the best faculty and the ability to release those who may not meet the rigorous standards.

An observation in the report inferred that CGSC used contracted faculty that had limited operational experience. Firstly, all faculty at CGSC are government employees, the vast majority hired as Title X, who go through a rigorous screening process. Secondly, the majority of the civilian faculty members at CGSC are retired military officers from all services, many of whom have extensive operational experience. Consequently, they possess a thorough knowledge of the Department of Defense and their particular service. However, recent operational experience is not a sufficient criterion for teaching at CGSC. This is a graduate school which employs an adult learning model highlighted by facilitated discussion in the classroom. The civilian faculty at CGSC is dedicated to becoming expert teachers and facilitators as well as subject matter experts and scholars in their chosen fields. It is the strong belief of the leadership at CGSC a mix of military and civilian faculty best achieves the graduate level educational goals of the College. Further, the augmentation of guest speakers provides operational currency as needed throughout the course.

Of course we would like to have a higher percentage of military faculty at ILE, but the requirements of the operational force have strained our ability to fill the generating force. The current ratio of 70 percent civilian and 30 percent military is a result of not having the requisite number of military faculty assigned to CGSC. A balance closer to 50% would allow us to better develop both the rotational military faculty and maintain closer ties to the operating force. As the operational environment changes, the availability of qualified military officers to teach at CGSC should improve; however, CGSC will not lower faculty standards to accept any military officer.

The quality of the facilities is an essential element of a quality education. The Army has invested in providing an optimum environment for our officers to learn. To accommodate the increase in International Fellows/Foreign Military Officers from 40 to 80 by AY2013, the Army War College is expanding from 20 to 24 classrooms, all of which will have extensive educational and informational technology. Office space will be increased to accommodate additional faculty. Both projects will be complete in June 2011. This expansion project also will reduce the number of students per seminar from 17 to 16, bringing it more in line with the adult learning model.

The CGSC Lewis and Clark Center is a state of the art learning center. It is four floors high, with 96 classrooms each configured with a horseshoe-shape of 16 workstations and an instructor station, white boards, plasma screens, video-teleconference capability, and cameras. Eisenhower Auditorium also features state-ofthe art technology and holds over 2,000 people.

CGSC views the integration of technology as critical to learning. The College has acquired and extensively leveraged appropriate technologies sufficient to enable the conduct of operational and tactical planning singularly or in concert with other units, service centers, and schools using battle command systems available in the field. Technology is fundamental to CGSC's mission and is a real strength of the institution.

CLOSING

We fully recognize and accept that we must continually adapt our institutions and policies to achieve a better balance of Professional Military Education as a part of the Army Leader Development Strategy. Today, the Army's challenge is to develop a sustainable learning environment that is essential to support operational adaptability. We are shaping the future of our PME system to better develop the leader characteristics our Expeditionary Army requires of its officers to lead Full Spectrum Operations in an era of persistent conflict. We must continue to prepare our leaders to be adaptive and open minded in framing the problems they will face, agile in considering new ideas, and grounded in the values to which our Nation aspires.

We look forward to continuing to work with the Subcommittee on these issues and related PME issues you have identified for further study. We recognize that your continued support of Professional Military Education programs is vital to the sustained health of our leader development and the security of our Nation. Furthermore, your continued concern and support of our Soldiers, Families, Civilians and Veterans who serve our great Nation is greatly appreciated.

In: Professional Military Education
Editor: Michelle P. Jacobs

ISBN: 978-1-62948-814-1

Chapter 3

STATEMENT OF SCOTT LUTTERLOH, DIRECTOR, TOTAL FORCE TRAINING AND EDUCATION DIVISION, U.S. NAVY. HEARING ON "CONTINUED ENGAGEMENT: DOD RESPONSES TO THE HASC REPORT ON PROFESSIONAL MILITARY EDUCATION"*

Chairman Snyder, Representative Wittman, and distinguished members of the Oversight and Investigations Subcommittee, thank you the opportunity to discuss Navy's views on the Committee's April 2010 report, *Another Crossroads? Professional Military Education Two Decades after the Goldwater-Nichols Act and the Skelton Panel.*

We appreciate the Subcommittees efforts in conducting such a comprehensive assessment of Professional Military Education (PME) since enactment of Goldwater-Nichols. The major findings are accurate in identifying the fundamental issues warranting critical deliberation as we continue in our commitment to improve PME and the professional development of the officer corps. We concur with the Subcommittee's assessment that, while the PME system is basically sound, there are areas that need improvement. We are engaged with the Joint Staff and other Services in a deliberate, collaborative assessment to evaluate recommendations to improve officer in-resident PME. Our approach is to seek solutions that enable development of a broad portfolio of core capabilities to support vital interests, realized by well-prepared, highly-motivated and ably-led professionals.

Navy places significant value on PME as we develop and enable resilient and adaptable leaders to meet challenges at the tactical, operational, and strategic levels of war. Naval War College, renowned for its PME program, has once again received accreditation through the Process for Accreditation of Joint Education (PAJE), for Joint Professional Military Education (JPME) Phase I, and first-time accreditation for JPME Phase II. Navy continues to emphasize PME as we provide unique and complementary maritime war-fighting skills to joint and combined force commanders. We support these commanders with operational forces led by commanding officers who have completed PME.

* This is an edited, reformatted and augmented version of a statement, presented November 30, 2010 before the House Armed Services Committee, Oversight and Investigations Subcommittee.

In response to the need for increased joint and service-specific subject matter to be taught earlier in an officer's career, Navy has established a career continuum of PME. We have a sequenced continuum of learning that provides relevant education aligned to career progression spanning E-1 through O-8, with a goal of providing Navy's Total Force with a standardized, comprehensive understanding of the Navy and its war-fighting capabilities. PME, at the primary, intermediate, senior, and flag levels, involves developing habits of thought and key attributes such as critical thinking, effective communication, and risk management.

We concur with the report's observation that, while PME is a factor in cultivating strategists, it is not the primary means for developing future strategic decision-makers. As noted, Navy has a relatively advanced process for cultivating strategists. We are actively engaged in a review of how we develop our senior leaders, as we work to further enhance development and tracking of strategists, through improvements to Navy's politico-military education, utilization and community sponsorship program.

Competing demand for time in the career track of officers of the Unrestricted Line (URL) remains a primary challenge. We continually review officer development timelines to explore innovative avenues to provide increased opportunities for education, training and experience. Recently, we implemented new approaches to officer development through introduction of specialty career paths for URL officers, which are not typically available within the warfare enterprises. These specialty career paths, including Antiterrorism/Force Protection, Missile Defense, Strategic Sealift, Operations Analysis, Operational Planner and Training Management, allow Navy to better integrate training, education and experience tours focused on specialty areas, while officers continue to serve in their warfare communities. Our executive officer/commanding officer (XO/CO) fleet-up program, by which prospective CO's are first assigned to serve as XO of the ship or squadron they will ultimately command, capitalizes on direct experience received within a particular command, and captures valuable career path time for fulfilling other developmental assignments, including education. We are continuing to explore other potential innovative opportunities.

We concur with the report's observation that Naval War College's detailed appointment letters provide well-defined expectations related to appointment renewals and performance appraisals, and serve as positive examples for use by the other services to recruit and sustain quality faculty. We are currently evaluating the report's recommendation that Navy consider instituting a quality board process for selection of in-resident PME students, by evaluating our process of screening top performing officers for eligibility to attend service colleges. This recommendation was an agenda item before the Vice Chief of Naval Operations-chaired Advanced Education Review Board (AERB) last month. Navy has appointed a cross-functional working group to evaluate the current selection board processes for JPME in-residence education. Results of the group's evaluation will be presented to the AERB in April 2011.

Navy takes a balanced approach to professional education that views operational competency and primacy of command as the key professional measures for Naval Officers. PME has been instrumental in developing a highly educated and more effective leader. We value the flexibility provided by the Chairman of the Joint Chiefs of Staff, which permits Navy to manage the content, quality and conduct of our PME Continuum.

Thank you again for your strong support of military members and their families and for your leadership in the professional development of the Navy Total Force.

In: Professional Military Education
Editor: Michelle P. Jacobs
ISBN: 978-1-62948-814-1

Chapter 4

STATEMENT OF DANIEL R. SITTERLY, DIRECTOR OF FORCE DEVELOPMENT, U.S. AIR FORCE. HEARING ON "CONTINUED ENGAGEMENT: DOD RESPONSES TO THE HASC REPORT ON PROFESSIONAL MILITARY EDUCATION"*

INTRODUCTION

The Chief of Staff of the Air Force (CSAF) concurs with the general tenets of the House Armed Services Committee report that the "Professional Military Education system is still basically sound" but with systemic and institutional areas that require our heightened focus and effort to improve. We strongly support the idea that education must be relevant to the student and the Services as well as the Department of Defense. To maintain a stronger correlation among education, officer development, and organizational requirements, the Air Force will ensure Professional Military Education (PME) can anticipate and adapt to current and future challenges, not the least of which is meeting the increasing need for additional joint and Service-specific subject matter taught earlier in an officers' career, and equipping officers with the developmental joint education they need before they are placed into joint assignments.

Joint PME

Falling under the broad genre of system issues, we concur with the need to address the diminished relationship between Joint PME and joint duty assignments. The Air Force is a force provider and, as such, endeavors to meet the requirements detailed in both DoD and joint policy and directives. Addressing this issue requires both the individual and combined efforts of the Services. This same effort will be expended to address two related areas of

* This is an edited, reformatted and augmented version of a statement, presented November 30, 2010 before the House Armed Services Committee, Oversight and Investigations Subcommittee.

concern: The preparatory value of PME and tensions among career development, and joint-service requirements and PME.

Developing Strategists

Despite several subcommittee findings concerning our failure to identify and cultivate strategic thinkers, we are actively developing and implementing programs to grow strategists who think both broadly and critically. Combined with a strong Fellowship program, we have recently added a number of advanced academic degree opportunities in history, political science, international relations, economics, and philosophy to our portfolio. These new degrees are being earned at some of the nation's most prestigious universities and a number are reserved for only our captains. Offering these opportunities earlier in our officers' careers will help the Air Force develop leaders better prepared to meet current and emerging needs. Beyond that, a complete transformation of the company grade officer-level PME programs was just completed, offering an expanded introduction to the Service-specific doctrine in the Air and Space Basic Course, and a dramatically enhanced leadership course that includes joint and staff training at the Squadron Officer School. Through an increased focus on critical thinking at junior levels, we are developing an officer corps that is both capable of, and empowered to solve the problems they will encounter throughout their careers. They are learning and practicing strategic thinking earlier than in the past.

Those that stand out in their strategic-thinking capability have additional opportunities for development in the advanced-learning programs for exceptional students that we have created at both Air Command and Staff College and Air War College. Additionally, we have expanded the student throughput at our School of Advanced Air and Space Studies. We will continue to mentor and track these officers to ensure that their talents and additional specialized developmental education are put to good use in support of the Department's wide array of missions. Additionally, we are in the process of adding a doctoral program for senior development in the School of Advanced Air and Space Studies to extend exceptional students' intellectual growth beyond their residence experience. We are also working to ensure that our assignment policies reinforce the relationship between Joint PME and joint duty assignments, in order to provide officers who are grounded in the complexities of joint, coalition and multinational operations and, therefore, better prepared to overcome the unexpected challenges they will face.

Content and Delivery

Although the Air Force can address issues related to the Process for Accreditation of Joint Education (e.g., reviewing our selection process), the combined Services may need to consider changes in the policy governing student-mix ratios due to faculty manning as a result of operations tempo and end strength. Another issue falling under the heading of content and delivery is our continuing use of pass/fail systems. The Air Force concurs with the report's finding that these systems do not necessarily detract from the rigor of academic programs; however, the Air Force believes Air University's current online distance learning program serves as the model for design, allowing us to evolve beyond the more simplistic correspondence-course approaches of the past. We also concur with the report's finding that we have a need to address faculty and resource concerns at our institutions. Therefore, we are reviewing policies regarding our hiring practices, job advancement and academic freedom, as well as copyright and intellectual property concerns.

CONCLUSION

The Air Force has made significant advances in the past two years toward its goal of leading the Services in its approach to developmental education. We have "upped our game" with a new distance learning academic/PME program, the complete transformation of company-grade officer Professional Military Education, new advanced courses within the Air Command and Staff College and Air War College, and expanded enrollment and progress toward a doctoral program in the School of Advanced Air and Space Studies. These efforts, combined with conducive personnel policy and developmental opportunities allow us to continue forging the synergistic relationship between Air Force training, experience, and education. We're confident we can continue to produce the officers our nation needs, and deliver them to their assignments fully trained, equipped and ready to excel.

In: Professional Military Education
Editor: Michelle P. Jacobs

ISBN: 978-1-62948-814-1

Chapter 5

STATEMENT OF MAJOR GENERAL ROBERT NELLER, PRESIDENT, MARINE CORPS UNIVERSITY. HEARING ON "CONTINUED ENGAGEMENT: DOD RESPONSES TO THE HASC REPORT ON PROFESSIONAL MILITARY EDUCATION"*

Chairman Snyder, Ranking Member Wittman, distinguished sub-committee members. Good morning. Thank you for permitting me to discuss the Marine Corps' views on *Another Crossroads? Professional Military Education Two Decades after the Goldwater-Nichols Act and the Skelton Panel.*

The Marine Corps deeply appreciates the extensive research, analysis, and documentation contained in the report. The work of the subcommittee provided a thorough assessment of the status of Professional Military Education (PME) and more importantly, provided sound recommendations on a variety of issues that will improve our educational programs. We have carefully reviewed the report and are already at work implementing many of the recommendations. Additionally, we are working closely with the Joint Staff and other services to ensure a coordinated approach as we examine each issue.

The Marine Corps, and Marine Corps University (MCU) in particular, is constantly reviewing and revising PME programs to ensure we meet the needs of the operating forces and prepare our leaders for the complex environments in which they will operate. Over the past few years, we have strengthened the faculty and staff of our schools and colleges and continually reviewed our curricula for relevance. Additionally, we have planned infrastructure improvements and technology enhancements that will dramatically improve the learning environment for our students. While we continue to make our quality resident officer programs even better, our current emphasis is the improvement of our enlisted PME programs. We ask an unbelievable amount from our enlisted leaders and they deserve the same quality education as our officers. We are actively revising our enlisted curricula to ensure relevance at all levels from corporals to senior enlisted leaders. Online distance education modules are being developed to both better reach Marines participating in PME

* This is an edited, reformatted and augmented version of a statement, presented November 30, 2010 before the House Armed Services Committee, Oversight and Investigations Subcommittee.

programs, and to provide a higher quality educational experience. These distance programs will ultimately serve as a prerequisite for resident courses. This will allow the resident programs to focus on the application of warfighting skills, critical thinking, and communication skills. The recent implementation of our Senior Enlisted PME Course fills a critical void in the enlisted PME continuum by enhancing the knowledge of our senior leaders to allow them to more effectively serve as senior advisors. As noted in the *Issues for Further Study* section of the report, progress has been made but much work remains to fully prepare our enlisted leaders.

Last month, General Amos published his Commandant's Planning Guidance identifying professional military education as one of his top priorities. His guidance directs the development of plans to increase the number of Marines attending resident officer and enlisted education programs and to continue to develop MCU into a world-class institution. We are in the process of developing options to increase attendance in our programs without sacrificing quality or desired learning outcomes. World-class status is achieved through world-class faculty, world-class students, world-class curricula, and world-class facilities and resources. Although work remains, as noted earlier, we have made substantial progress in strengthening faculty, students, and curricula. We are also on track to make significant progress in facilities. For example, over $120M in military construction is programmed over the next three years for our educational facilities. I am also pleased to report that the Expeditionary Warfare School Distributed Education Network (EDEN), an item of interest during a subcommittee visit, is now fully funded and will be operational as soon as possible. Our institutional commitment toward PME, coupled with the Commandant's commitment to achieve world-class status, illustrate the Marine Corps is totally dedicated to educating our future leaders. The Commandant's Planning Guidance and the subcommittee report complement each other and provide a good roadmap to improve our already strong PME programs.

I will address some of the key recommendations of the subcommittee's report, particularly those that relate directly to the Marine Corps.

The report correctly identifies the need to clarify the role of education in preparing officers for joint duty assignments. We agree that PME, and Joint PME (JPME) in particular, plays an important role in preparing officers for these challenging joint assignments. We further believe that a combination of training and education is needed to fully prepare officers. Surveys of supervisors of our graduates indicate that our programs are meeting this goal. In fact, for the past two years, 98% of the supervisors indicated that Command and Staff graduates are prepared for joint assignments, while 100% of supervisors indicated that MCWAR graduates are prepared for joint assignments. Surveys of the graduates of both programs were equally positive. Additionally, our educational programs provide comprehensive coverage of the competencies identified in the report as appropriate for educational venues in the development of joint staff officers. Service capabilities, the ability to conduct research, critical thinking, problem solving, and effective communication skills are emphasized in all our colleges and schools. Over the past five years, MCU has continually raised the bar for effective communications. Most recently, MCU identified "Strengthening Leadership through Enhancement of Communication Skills" as a key effort for student learning and for submission as a component for regional accreditation. Under this program, MCU has established a Leadership Communication Skills Center, standardized assessments, and provided individualized communication skills assistance to numerous students. We view these skills as crucial for all officers, regardless of a particular billet assignment.

We agree that development of officers must be approached in a holistic fashion. Education is a critical component, but varied assignments, unit training, and individual developmental opportunities all ultimately contribute to the maturation of our leaders. MCU works closely with Manpower and Reserve Affairs to ensure PME programs are aligned with those of the manpower community. Ongoing planning to substantially increase attendance at officer resident PME programs will require close coordination with manpower to meet the Commandant's guidance, while maintaining adequate levels of manning in our operating forces. As suggested in the committee report, we are investigating innovative approaches to deliver PME in conjunction with operational assignments and other developmental opportunities.

The report correctly notes that the Marine Corps is the only service that does not have a formal program that allows Marines to pursue a doctoral degree. The Marine Corps is currently conducting a comprehensive education study to identify positions requiring specific educational degrees. This study will validate all existing structure positions, but more importantly, will identify opportunities for Marines to pursue degrees in a wide variety of disciplines. Doctoral degrees will be a key component of this study. MCU has proposed to the institution a program whereby selected Marines can attend civilian institutions to obtain a doctoral degree. Under this proposal, the officer will spend 18-24 months at a civilian institution to complete course work and comprehensive examinations, and then will be assigned as teaching faculty at MCU while completing his or her dissertation. The program will benefit the officer and the Marine Corps. We value the liberal arts education provided by civilian institutions and, as noted in the report, do not view internal-administered programs as substitutes for these developmental opportunities.

We concur with the report's observation that strategists are developed through a combination of experience and education. We believe that our PME programs are properly constructed to provide a thorough understanding of the principles of the strategic level of war and the complexities of developing well-reasoned strategy. In particular, our School of Advanced Warfighting provides a world-class curriculum for complex problem solving and planning and the Marine Corps War College is focused on development of strategic leaders. The previously mentioned graduate education study will identify positions that will provide unique developmental opportunities for Marines of all ranks. Additionally, we recently conducted a pilot Senior Planners Course for Colonels to hone their planning skills as they assume key positions in the operating forces. The results of the pilot were very positive and we plan to continue to offer the course to better prepare many of our senior leaders.

We understand the rationale of the report's recommendation that general criteria or credential requirements should be developed for leaders of PME institutions. While these would be useful in vetting candidates, we believe that demonstrated leadership performance should be the basis for selection of individuals to lead our PME institutions. Academic credentials are important, but the ability to lead large complex organizations, completion of PME at each level, and a concern for properly educating leaders are the key factors in selecting these leaders.

We appreciate the support of the subcommittee and we will review the associated recommendation to amend the 2007 legislation so as not to preclude the possibility of Joint Duty Assignment credit for non-host faculty at JPME I institutions. This is an important issue for our intermediate level service institutions as we seek to attract the most qualified personnel as faculty. The window of opportunity, however, is narrow for these officers as they must balance command, staff assignments, professional developmental opportunities, and joint duty assignments. These faculty members continually work joint matters such as national security strategy, campaigning, joint planning processes, joint capabilities, and the integration of service capabilities. Even though officers may self-nominate for the awarding of joint credit, this is far from a guarantee. We wholeheartedly agree with the subcommittee's recommendation to review the provisions of faculty hiring under Title 10 authority. The current statute has served us very well as we have been able to hire exceptional faculty for our 10-month officer PME programs. As mentioned previously, our enlisted leaders deserve the same quality educational experience. We view this as a key step in improving our enlisted programs and more closely aligning them with officer PME.

In summary, your Marine Corps Professional Military Education programs are strong and are producing leaders who are competent, able to think critically, and capable of solving complex problems. We appreciate the work of the subcommittee and believe the recommendations contained in the report will make our programs even stronger.

INDEX

D

E

F

G

H

I

J

K

L

M

N

O

P

Q

R

S

T

U

V

W

Y